Lecture Notes in Mathematics

Edited by A. Dold and B. Eckmann

690

William J. J. Rey

Robust Statistical Methods

Springer-Verlag
Berlin Heidelberg New York 1978

Author
William J. J. Rey
MBLE – Research Laboratory
2, Avenue van Becelaere
B-1170 Brussels

Library of Congress Cataloging in Publication Data

Rey, William J. J. 1940-
Robust statistical methods.

(Lecture notes in mathematics ; 690)
Bibliography: p.
Includes indexes.
1. Robust statistics. 2. Nonparametric statistics.
3. Estimation theory. I. Title. II. Series:
Lecture notes in mathematics (Berlin) ; 690.
QA3.L28 no. 690 [QA276] 510'.8s [519.5] 78-24262

AMS Subject Classifications (1970): Primary: 62G35
Secondary: 62G25, 62J05

ISBN 3-540-09091-6 Springer-Verlag Berlin Heidelberg New York
ISBN 0-387-09091-6 Springer-Verlag New York Heidelberg Berlin

Printed in Germany

Printing and binding: Beltz Offsetdruck, Hemsbach/Bergstr.
2141/3140-543210

FOREWORD

During the last nine years, several problems in the statistical processing of biomedical data have been encountered by the author. These problems had in common the fact that most of the usual assumptions were without any solid basis. Poor quality samples drawn from unknown distributions, usually non-normal and frequently non-stationary, were the ordinary lot; nevertheless, sophisticated parameters had to be reliably estimated and the scatter of the estimators was needed to permit comparison of the results. This has been partly solved by application of robust methods.

The methods presented in this text are oriented toward the design of robust estimators. The primary concern is preventing any significant offset of the estimates due to the selection of an erroneous model or to spurious data in the sample. The second concern is bias reduction and variance estimation. The special emphasis reserved to type M estimators is justified by their analytical form which permits to assess their properties, even for small sample sizes ($n=10$ or 50), without resorting to involved arguments.

The theoretical tools are mainly the jackknife and the influence function. Applied derivations are in the fields of location estimation and regression analysis. Due attention is devoted to computational aspects.

TABLE OF CONTENTS

1. INTRODUCTION

The term "robustness" does not lend itself to a clearcut statistical definition. It seems to have been introduced by G.E.P. Box in 1953 to cover a rather vague concept described in the following way by Kendall and Buckland (1971). Their dictionary states :

> "Robustness : Many test procedures involving probability levels depend for their exactitude on assumptions concerning the generating mechanism, e.g. that the parent variation is Normal (Gaussian). If the inferences are little affected by departure from those assumptions, e.g. if the significance points of a test vary little if the population departs quite substantially from the normality, the tests on the inferences are said to be robust. In a rather more general sense, a statistical procedure is described as robust if it is not very sensitive to departure from the assumptions on which it depends."

This quotation clearly associates robustness with applicability of the various statistical procedures. The two complementary questions met with can be expressed as follows : first, how large is the domain of applicability of a given statistical procedure or, equivalently, is it robust against some departure from the assumptions ? Second, how should we design a statistical procedure to be robust or, in other terms, to remain safe in spite of possible uncertainty in the available data set ?

1.1. History and main contributions

Due to the appearance of involved analytical as well as computational facilities, the field of robustness has received an increased attention during the last thirty years. Mainly, progresses in non-linear mathematics and in recursive algorithms have permitted the new developments. However, it stands in the line of many old studies. For instance, a mode can possibly be seen as a robust estimate of location, as it has been some twenty four centuries ago. Thucydides relates :

> "During the same winter (428 B.C.), the Plataeans ... and the Athenians who were besieged with them planned to leave the city and climb over the enemy's walls in the hope that they might be able to force a passage ... They made ladders equal in height to the enemy's wall, getting the measure by counting the layers of bricks at a point where the enemy's wall on the side facing Plataea happened not to have been whitewashed. Many counted the layers at the same time, and *while some were sure to make a mistake, the majority were likely to hit the true count*, especially since they counted time and again, and, besides, were at no great distance, and the part of the wall they wished to see was easily visible. The measurement of the ladders, then, they got at in this way, reckoning the measure form the thickness of the bricks."

Eisenhart (1971) has also compiled other examples of more or less sophisticated procedures to estimate a location parameter for a set of measurements.

Historical background can also be gained by the account of the state of affairs around 1900 given in Stigler (1973). Briefly, it appears that the adoption of least squares techniques was seen as second best to unmanageable approaches. The dogma of normality was largely accepted; that is, observations in disagreement with the dogma were seen as erroneous and had to be discarded. At that time, an important effort was done in the investigation of various rejection procedures as well as in the assessment of their soundness. A remarkable report of this history up to the present time can be found in the series of papers written by Harter (1974-1976). With regard more specifically to robustness history, motivations and theoretical developments are surveyed in the 1972 Wald Lecture of Huber (1972-1973) and complementarily by Hampel (1973). We will later join them.

To gain some more insight in what is robustness we now sketch some important works of the recent past :

- The investigations of von Mises (1947) and Prokhorov (1956) have clarified the relations which exist between asymptotic theory of estimation and finite size practical situations. They provide a neat framework where the theory of robustness has been able to grow. Particularly, they provide justifications for the analytical derivations.

- A mathematical trick proposed by Quenouille (1956) and augmented by Tukey (1958), the jackknife technique, permits to reduce the bias and estimate the variance of most estimators without any regard for the distribution underlying the data set. The statistician is thus released of some frequently questionable distribution assumptions.

- A review of the principles pertaining to the rejection of outliers by Anscombe (1960) has stimulated theoretical and experimental researches on how to take into account observations appearing in the tails of the sample distributions.

- The famous paper of Huber, in 1964, has been at the origin of answers to the question on how to design robust statistical procedures. He considers domains of distributions; we excerpt : "A convenient measure of robustness for asymptotically normal estimators seems to be the supremum of the asymptotic variance ($n \to \infty$) when F (the distribution of the sample) ranges over some suitable set of underlying distributions". Further, he introduces M-estimators and characterizes a most robust family among them for the estimation of the location, when the underlying distribution is a contaminated normal. In this framework, we find the estimators obtained through minimization of some power p of residuals; the classical least squares estimations ($p = 2$) have here their places, the pioneer work of Gentleman (1965) is referred to for other p-values. The maximum likelihood estimators also are M-estimators.

- A thesis by Hampel (1968), related to von Mises' work, introduces the influence curve as a tool exhibiting the sensitivity of an estimator to the observation values. Thus, it permit to modify estimation procedures in order not to depend on outliers, or on any other specific feature of the observations.

- As far as the literature of the period prior to 1970 is concerned, we refer the interested reader to the annotated bibliography prepared under arrangement with the U.S. National Center for Health Statistics, (N.C.H.S. - 1972).

- To conclude the list of main theoretical contributions, it may be worthy to mention that many fundamental questions remain open. Among them, the question of what we really estimate is approached by Jaeckel (1971), and, whether estimators are admissible is analysed by Berger (1976a, 1976b).

- Complementary to the theoretical progress, experience has been gained through Monte Carlo computer runs. In this respect, the Princeton Study (Andrews et al. - 1972) specially deserves mention; it has displayed in an obvious way the need for robust statistical procedures.

1.2. Why robust estimations ?

It seems to this author that the main motivation for making use of robust statistical methods lies in the overwhelming power of our computing facilities. In short, it is so easy to perform statistical analysis with computers that, rather frequently, data sets are processed by unsuitable softwares. Comparison with results produced by robust methods then draws attention on possible deficiencies of the data sets as well as on limitations of the applied statistical procedures. Thus, it appears that robustness is essential because it is complementary to the classical statistics. One and the other must contribute to the elaboration and to the validation of statistical conclusions.

Many statisticians do not know how poor can be the methods they apply when the data sets do not strictly satisfy the assumptions. This has been brillantly illustrated by Tukey (1960) in his study of contaminated normal distributions. We have further extended this study - see entry "contaminated normal distribution" of the appendix - and arrive at the conclusion that quite large sample sizes are required to justify the measurement of the scale by the standard deviation rather than by the mean deviation. Some eight thousand observations should be at disposal to guarantee (at level 0.95) that the standard deviation is the most efficient estimator. This fact must be kept in mind.

In a rather incentive paper some justifications to the use of robustness have been layed down by Hampel (1973). Hereunder, we propose a long excerpt :

"What do those "robust estimators" intend ? Should we give up our familiar and simple models, such as our beautiful analysis of variance, our powerful regression, or our high-reaching covariance matrices in multivariate statistics ? The answer is no; but it may well be advantageous to modify them slightly. In fact, good practical statisticians have done such modifications all along in an informal way; we now only start to have a theory about them. Some likely advantages of such a formalization are a better intuitive insight into these modifications, improved applied methods (even routine methods, for some aspects), and the chance of having pure mathematicians contribute something to the problem. Possible disadvantages may arise along the usual transformations of a theory when it is understood less and less by more and more people. Dogmatists, who insisted on the use of "optimal" or "admissible" procedures as long as mathematical theories contained no other criteria, may now be going to insist on "optimal robust" or "admissible robust" estimation or testing. Those who habitually try to lie with statistics, rather than seek for thruth, may claim even more degrees of freedom for their wicked doings.

"Now what are the reasons for using robust procedures ? There are mainly two observations which combined give an answer. Often in statistics one is using a parametric model implying a very limited set of probability distributions thought possible, such as the common model of normally distributed errors, or that of exponentially distributed observations. Classical (parametric) statistics derives results under the assumption that these models were strictly true. However, apart from some simple discrete models perhaps, such models are never exactly true. We may try to distinguish three main reasons for the deviations : (i) rounding and grouping and other "local inaccuracies"; (ii) the occurrence of "gross errors" such as blunders in measuring, wrong decimal points, errors in copying, inadvertent measurement of a member of a different population, or just "something went wrong"; (iii) the model may have been conceived only as an approximation anyway, e.g. by virtue of the central limit theorem".

1.3. Summary

It appears that most of the robust methods have a weak basis; this fact must be faced and can be related partly to some conceptual difficulty. When those methods are required, it is due to some uncertainty concerning the statistical model, the quality of the sample at disposal as well as, perhaps, uncertainty concerning the issues involved (e.g., selection of an optimal loss function). These various factors prohibit any clear definition of the parameter to estimate and, thus suppress any possibility of comparing an estimator with the parameter. In practice, all estimators will be analysed by reference to their asymptotic values obtained for infinite sample sizes. Furthermore, very often it is possible to compare estimators under given model at the asymptotic level; therefrom, we will encounter robust estimators which are consistent to some parameter. However what is estimated when the model assumption is erroneous may be unclear. This is the conceptual difficulty and it appears at the root of all robustness principles.

Section 2 provides the few theoretical derivations which permit to compare an estimator to its asymptotic value. The result has the form of a Taylor-like expansion; an argument is presented to validate the obtained structure, but a strict demonstration has not been possible in spite of several attempts during the last thirty years. We may however conjecture that the needed conditions of validity are scarcely restrictive. This theory directly leads to a definition of robustness as well as to the influence function.

Section 3 provides reduction of possible bias in estimators as well as estimation of the variance (or covariance) through the jackknife method. To the best of our knowledge, the presented derivations are original and permit to substantiate several conjectural results.

Section 4 is the central part of this text. It is a detailled analysis of M-estimators and of simultaneous M-estimators (denoted MM), with emphasis on applications in regression problems. This part presents several generalizations of previous developments and is original in many respects.

Section 5 reconsiders a few specific questions which have been previously met, but left on the side because they were not timely. Most of the questions are, so far, open.

Section 6 is a limited bibliography. We have tried to favor recent papers and surveys in the domains which are secondary to our main investigation. We are well aware of the fact that such selection is very arbitrary. Furthermore, we feel limited in readability, time and space.

Throughout this text, we have not devoted much attention to the asymptotic properties and have placed the accent on the behaviour of finite sample estimators. This option results from the need of robust methods in applications.

2. ON SAMPLING DISTRIBUTIONS

2.1. Scope of the section

Hereinafter, we intend to sketch two theories relative to the relationships existing between distributions for one of them, and relative to the relations between distributions and estimators for the other.

To fix the ideas, consider the following standard set of equations

$$\begin{aligned} m &= \sum w_i\, x_i, \qquad (i=1,\dots,n), \\ \sum w_i &= 1, \\ s^2 &= \sum w_i (x_i - m)^2, \\ \mu &= \lim m,\ n \to \infty, \\ \sigma^2 &= \lim s^2,\ n \to \infty. \end{aligned}$$

Under appropriate conditions, it can be used to estimate the location μ of the one dimension sample $(x_1,\dots,x_n)$, as well as its scale σ. In the present formulation each observation x_i has been attributed a relative weight w_i, which could be reflecting its importance or its accuracy.

To investigate the dependence of μ and σ on the distribution underlying the observations, we need a tool to compare distributions such that "closeness" between distributions lead to "closeness" of estimators. This will be our first concern.

The interest will also lie in the sample distribution of m and s or, more specifically, in the dependence with respect to each observation x_i, to the relative weight w_i as well as to the sample size n. In the present case, the relations are enlightened by expansions in terms of the asymptotic estimators; the scale parameter is also given by

$$s^2 = \sigma^2 + \sum w_i\,[(x_i-\mu)^2 - \sigma^2] - \sum\sum w_i w_j\,(x_i - \mu)\,(x_j - \mu),$$

which lends itself to easy analysis. This will be our second concern.

2.2. Metrics for probability distributions

As indicated by Munster (1974), we do not have to restrict our distributions to Baire functions, nor the application space to Borel sets; although, in practice, we only feel at ease with the Baire class of functions and the Euclidean space R^n. The fundamental discontinuous distribution we make use of is the classical Dirac function defined over the n-dimension space R^n - Complementary precisions are given in the appendix at the section on "distribution space" - We will be only concerned here by ways of measuring how close distributions may be to one another; the assessment of closeness between estimators will later be reduced to the closeness of their sampling distributions and, thus, only distributions are of interest.

A great number of "distance" definitions have been proposed to assess the closeness of distributions, however they are generally not acceptable in our context. Going through the review of Kanal (1974) or with the help of the paper of Chen (1976), the following deficiencies are observed for most distances :

- They provide a measure of the closeness of continuous distributions, but, are not appropriate to compare an empirical (discrete) distribution with its underlying parent.
- They may be strictly one-dimension and, thus, are not applicable on R^n.
- They rarely satisfy the triangular inequality, a very helpful condition to compare an empirical distribution with a continuous distribution differing from its parent.

To the best of our knowledge the only distance measure suitable to our context has been proposed by Prokhorov (sec. 1.4, 1956), it is simple in its main idea although rather involved in its details - see "Prokhorov metric" in the appendix - To help in the understanding of its analytics, we now consider the three points just mentioned here above. The Prokhorov metric permits the comparison of a discrete empirical distribution with a continuous one through the association of each observation of the former with a subset of the sample space; the comparison is then performed with the help of the probability of the latter distribution over this subset. The distance measure is derived from the probability measures achieved over subsets of the sample space and is, therefore, independent of dimensionality considerations. In fact, its definition is relative to supremum of probability measures and, accordingly, the triangular inequality holds true.

We only need incidentally the Prokhorov metric definition and will accordingly limit ourselves to the present level of information. We conclude, in this respect, by an illustration.

We estimate the distance between two distributions defined by the densities of probability f and g over some n-dimension space. Distribution g is derived from f through

$$g(x) = (1-t)\ f(x) + t\ \delta(x-x_0), \qquad 0 \leq t < 1$$

where x is any point of the sample space and $\delta(x-x_0)$ is a Dirac function centered on x_0, i.e.

$$\delta(x-x_0) = 0, \quad \text{if } x \neq x_0$$
$$\int \delta(x-x_0)dx = 1$$

with the integration variable spanning the whole R^n space. For simplicity, we assume that f(x) satisfies the mean value theorem in x_0 that is

$$F(r) = \int_{|x-x_0|<r} f(x)\ dx = f(x_0)\ V(r)$$

where

$$V(r) = [\pi^{n/2}/\Gamma(1+n/2)]\ r^n$$

is the volume of the ball of radius r centered in x_0. Then the distance between f and g is given by

$$d(f,g) = \sup_r \inf_\varepsilon \{\varepsilon : G(r) \leq F(r+\varepsilon) + \varepsilon\}.$$

The inequality can be omitted when f(x) is sufficiently regular, therefrom ε may be seen as a function of r in

$$G(r) = F(r+\varepsilon) + \varepsilon$$

or

$$t + (1-t)\ f(x_0)\ V(r) = f(x_0)\ V(r+\varepsilon) + \varepsilon$$

and the distance is given by the maximum ε-value for all positive r-values. This supremum is realized by r=0. We obtain the recursive solution

$$d(f,g) = t - f(x_0)\ V[d(f,g)]$$

where the last term is neglegible with small t and n > 1. In quite general terms we observe that

$$d(f,g) < t \leq 1;$$

but we must warn the reader : Evaluation of this Prokhorov distances frequently requires fairly involved arguments.

2.3. Definition of robustness, breakdown-point

This paragraph essentially restates in other words the viewpoint of Hampel (1971) - For technical details, the reader is referred to the appendix.

We consider a set of observations $\{x_1,\ldots,x_n\}$ drawn from some distribution $f(x)$; this set will be used to estimate some parameter θ, let θ_n be the estimate. The sampling distribution of this estimate is noted $\phi(\theta_n,f)$ and depends upon $f(x)$. But generally we do not know $f(x)$ and only have a more or less valid model, say $g(x)$. We say in coarse terms that θ_n is robust if it is scarcely dependent upon the difference between $f(x)$ and $g(x)$; that is we expect $\phi(\theta_n,f)$ and $\phi(\theta_n,g)$ to be close.

More precisely, θ_n is said to be robust with respect to distribution g (and to f) if

$$d(f,g) < \delta \Rightarrow d[\,\phi(\theta_n,f),\ \phi(\theta_n,g)] < \varepsilon$$

for small positive ε and δ. It may happen that small ε be allowed although δ is not so small but remains inferior to some critical value δ^*. This value δ^* is the so-called breakdown-point. If $f(x)$ and $g(x)$ are not close enough (differ more than δ^*), the estimates of θ_n based respectively on $f(x)$ and on $g(x)$ may be quite different. Then, θ_n is not anymore robust.

The limited realism of these definitions must be indicated. The condition

$$d(f,g) < \delta$$

covers any distribution $f(x)$ in the δ-neighbourhood of $g(x)$ despite the fact that some $f(x)$, close to $g(x)$, can be not acceptable for extraneous reasons. Thus, it sometimes appears that a non-robust estimator is seen as robust for some subset of distributions $f(x)$, particularly when $f(x)$ is restricted to be member of some parametric family. Nevertheless, there exist non-robust estimators; a startling case is the mean, the ordinary arithmetical mean. Effectively, small contaminations (small δ) can yield any large offset of the mean (large $\varepsilon \leq 1$). In this case the breakdown-point is $\delta^* = 0$. On the contrary the median of a one-dimension distribution exhibits extreme robustness with $\delta^* = 1/2$. Up to a half of the sample can be outlying without leading to important error of estimation. An investigation of the breakdown-point utility has been reported by Hampel (1976) with special emphasis on location estimation with one-dimension samples.

To conclude let us add that the above definition of robustness concurs with the minimax approach of Huber (1964), particularly surveyed in a group of lectures (1969). The already mentioned idea is to design an estimator to be the "best" with respect to the least favorable distribution of a distribution subset.

2.4. Estimators seen as functionals of distributions

In a paper on the asymptotic properties of sample distributions, von Mises (1947) introduces in a heuristic way some Taylor-like expansion of estimators in terms of the distribution of the underlying sample. This expansion is valid in a not-clearly defined context of "differentiable statistical functions". Hereinafter we present this theory and provide an original delineation of the domain of application. But first, let us introduce the argument of von Mises.

The basic material is two distributions f and g, which can be viewed as points in a convenient distribution space, and an estimator, or rather a functional, on these distributions; let it be $T(f)$ or $T(g)$. The estimators are defined over some sample space we assume common to both and which could be discrete as well as continuous, thus far. We are now interested in the difference between the two functionals $T(f)$ and $T(g)$.

Consider the functional $T(h)$ where h is a distribution intermediate between f and g, precisely

$$h(x) = (1-t)\ f(x) + t\ g(x)$$

with

$$0 \leq t \leq 1.$$

This functional is, under suitable conditions, a continuous function of the real variable t and can be expanded in Taylor series, i.e.

$$T(h) = T(f) + t\,a_1 + (t^2/2)\,a_2 + \dots ,$$

which is converging inasmuch as T(f) and T(g) are finite. The coefficients a_i are given in terms of derivatives of T(h) with respect to the variable t and therefore involve the difference $[g(x)-f(x)]$. For instance, the coefficient a_1 comes out as

$$a_1 = \int u(x)\,[g(x) - f(x)]\,dx$$

or, equivalently,

$$a_1 = \int \psi(x)\,g(x)\,dx.$$

Before proceeding, we note that the closer h is from f, the smaller are the high order terms. Thus, when g and f are close, the expansion for t = 1

$$T(g) = T(f) + \int \psi(x)\,g(x)\,dx + \frac{1}{2}\int\int \psi(x,y)\,g(x)\,g(y)\,dx\,dy + \dots$$

can be truncated to its first few terms. The terms are the so-called "derivatives". They involve functions $\psi(x)$, $\psi(x,y)$, ... defined solely with the help of T(f) and f. They cancel for equal distributions g and f.

It has not appeared possible, so far, to state the conditions required to justify the above intuitive derivation. In this respect, von Mises satisfies himself by refering to previous works of Volterra, although they do not appear helpful in practice. Moreover, while a proper derivation is produced in order to substantiate the expansion, many conditions appear which can hardly be related in an easy way to the function T(f) and to the distributions f and g. The conditions involve the Prokhorov metric, bounding of some derivatives as well as existence and convergence of some integrals; in short, they involve features which are frequently unknown in practice. -See the appendix at the "von Mises derivative" entry- This state of affair being met with, we have preferred to limit ourselves to a set of sufficient conditions of general applicability - This is tentative.

We first define two distribution properties :

Definition 1. A distribution f is "smooth" with respect to functional T if, and only if, $\{T(f_i)\}$ converges to T(f) when $\{f_i(x)\}$ is any Cauchy sequence of continuous distributions uniformely converging to f(x).

Definition 2. A distribution f is "domain limited" with respect to functional T if, and only if,

$$\lim T(f_R) = T(f)$$

when R tends to infinity and $f_R(x)$ is the possibly defective distribution

$$f_R(x) = f(x), \quad \text{if } |x| \leq R$$
$$= 0 \quad , \quad \text{if } |x| > R.$$

It must be observed that these two definitions are constraining the functional T rather than the distribution f(x). They yield :

Theorem : A Taylor-like expansion in terms of the von Mises derivatives is valid for a functional T defined over two distributions f and g, when both distributions are smooth and domain limited with respect to the functional. For the real variable t satisfying

$$0 \leq t \leq 1,$$

we have

$$T[(1-t)f + tg] = T(f)$$
$$+ t \int \psi(x)\, g(x)\, dx$$
$$+ \frac{t^2}{2} \int \int \psi(x,y)\, g(x)\, g(y)\, dx\, dy$$
$$+ \ldots$$

To gain further insight in this theorem, we illustrate by the expansion of an estimator in term of its asymptotic value.

Let us say we have at disposal some sample $(x_1, \ldots, x_n)$ drawn from a p-dimension sample space $(p \geq 1)$ with continuous probability density function f(x). The sample population has possibly been stratified and, thus, we attribute to each observation a given positive weight; let them be noted $(w_1, \ldots, w_n)$. The empirical density functions is, accordingly,

$$g(x) = (1/\sum w_i) \sum w_i \; \delta(x-x_i), \qquad (i=1,\ldots,n)$$

where $\delta(x-x_i)$ is the Dirac function concentrated on the observation x_i. We now introduce the parameter θ defined by the functional T(f) and estimated by T(g), i.e.

$$\theta = T(f)$$
$$\hat{\theta} = T(g) = T(x_1,\ldots,x_n;\ w_1,\ldots,w_n);$$

we further assume that T(g) is analytical with respect to the observations x_i as well as to the weights w_i. With this restriction on the functional T, the distributions f and g are smooth and domain limited. Effectively, distribution f(x) is smooth per definition and domain limited per existence of θ, the parameter to be estimated. Distribution g(x) is smooth with respect to functional T, for $T(g^*)$ converges to T(g) with m tending to infinity in

$$g^*(x) = (1/\sum w_i) \sum w_i\ h_m(x-x_i)$$

and

$$\lim h_m(x) = \delta(x),\quad m \to \infty.$$

Any sequence of continuous functions $\{h_m(x)\}$ uniformely converging to the Dirac function may be considered. The last condition for the theorem applicability is that distribution g(x) be domain limited; it is inasmuch as it has been possible to observe the sample. Thus we have the expansion

$$\begin{aligned}\hat{\theta} &= T(g)\\ &= T[(1-t)\ f + tg],\quad \text{for } t = 1\\ &= \theta + (1/\sum w_i) \sum w_i\ \psi(x_i)\\ &\quad + \frac{1}{2} (1/\sum w_i)^2 \sum \sum w_i\ w_j\ \psi(x_i,x_j)\\ &\quad + \ldots\end{aligned}$$

with the higher order terms having neglegible importance when the sample is representative of the parent distribution f(x) or, in other terms, when f(x) and g(x) are close.

It may be noticed that the expansion given for the variance estimator in section 2.1 has precisely the above structure.

2.5. The influence function of Hampel

Under the appropriate conditions of regularity met in the above section, we see that an estimator $\hat{\theta}$ is related to the corresponding parameter θ by the simple approximate relation

$$\hat{\theta} = \theta + (1/\sum w_i) \sum w_i\ \psi(x_i)$$

where the factor $\psi(x_i)$ is indicative of the influence of the value x_i on the result $\hat{\theta}$. Hampel has named function $\psi(x)$, the influence function or, rather, the "influence curve" for he was considering only one-dimension samples. This is a very powerful tool to appreciate the robustness at one dimension - see Andrews et al. (1972) and Hampel (1974) - as well as at several dimensions - for instance, Rey (1975a) Basically it measures the sensitivity of $\hat{\theta}$ to each observation.

The influence function can also be defined, and easily obtained, as the first von Mises derivative in the direction of a Dirac function. Let us first recall the tautology

$$\int \psi(x)\ f(x)\ dx = 0.$$

Then, for the particular distribution

$$g(x) = (1-t)\ f(x) + t\ \delta(x-x_0)$$

we have, when t is small, that is when f(x) and g(x) are close to one another,

$$\begin{aligned} T(g) &= T(f) + \int \psi(x)\ g(x)\ dx \\ &= T(f) + t \int \psi(x)\ \delta(x-x_0)\ dx \\ &= T(f) + t\ \psi(x_0). \end{aligned}$$

Therefrom the following definition of the influence function occurs

$$\psi(x_0) = \lim \{[T(g) - T(f)]/t\},\ t \to 0.$$

Observe that the above definition has a larger domain of applicability than the Taylor-like expansion in terms of the von Mises derivatives; this is due to the very particular selection of distribution g(x) which involves only local properties at point x_0 of the sample space.

The concept of influence function can be immediately generalized to situations where several parent distributions are concerned as can be seen in Rey (1975b).

3. THE JACKKNIFE

3.1. Introduction

The so-called jackknife method has been introduced by Quenouille to reduce possible bias in estimation and then progressively extended to obtain estimation of variances. It is essentially interesting by its power to produce estimator improvement and estimator assessment in a cheap way, that is to say cheap in methodology but not necessarily cheap in computation if this has to be considered. By all standards, the results obtained are impressively good in most cases; but in a few not very well defined circumstances, the results are either poor or ridiculous.

Unfortunately the scope of application of the jackknife method is not easy to delineate. It may be seen, however, that it is applicable under suitable regularity of the estimator with respect to the observations. And, according to Huber (1972) : "It is hardly worthwhile to write down precise regularity conditions under which the jackknife has these useful properties, more work might be needed to check them than to devise more specific and better variance estimate" As indicated in the sequel, we disagree with the above viewpoint and support the use of the jackknife technique for all estimators falling in the frame of section 2.4.

But what is the jackknife method ? We now present the technique with relatively few details and with a relatively light notation used by Miller. More involved considerations could obscure the main ideas and will be reserved for the next section.

The story starts in 1956, when Quenouille proposes to reduce the possible bias of statistical estimators through what appears to be a "mathematical trick". He observes that an estimator $\hat{\theta}$ based on a sample of size n can frequently have its bias expanded in terms of the sample size as follows

$$E(\hat{\theta} - \theta) = a_1 n^{-1} + a_2 n^{-2} + \ldots$$

This form is now compared with the corresponding for the estimator $\hat{\theta}_i$ based on the sample of size n-1, the same sample without the i-th observation. The new expansion is

$$E(\hat{\theta}_i - \theta) = a_1(n-1)^{-1} + a_2(n-1)^{-2} + \ldots$$

when the observations are independent. To reduce the bias, Quenouille proposes to consider the variates

$$\tilde{\theta}_i = n\,\hat{\theta} - (n-1)\,\hat{\theta}_i$$

which have a similar bias expansion, except that the first order term cancels. It is

$$E(\tilde{\theta}_i - \theta) = -\,a_2/[\,n(n-1)] + \ldots$$

Obviously, the same mathematical trick can be applied to the second leading term of the expansion, and so on. To avoid a loss of efficiency in the estimation, he suggests the definition of an average estimator

$$\begin{aligned}\tilde{\theta} &= (1/n)\,\textstyle\sum_i \tilde{\theta}_i \\ &= n\,\hat{\theta} - [\,(n-1)/n]\,\textstyle\sum \hat{\theta}_i\,.\end{aligned}$$

For no clear reasons at that time, the jackknife estimate $\tilde{\theta}$ frequently demonstrates fairly good statistical properties. Tukey who will soon appear has also named $\hat{\theta}_i$, the pseudo-estimates, and $\tilde{\theta}_i$, the jackknife pseudo-values.

With regard to bias reduction it seems that it could possibly be advantageous to modify the sample size n by deletion of more than one observation, say we delete h observations ($h \geqslant 1$) and work out the pseudo-estimates with samples of size (n-h). There are many ways of constituting these subsamples and special consideration has been devoted to the three following schemes : first, deletion of ($h = 1$) observation at a time; second, deletion of ($h > 1$) consecutive observations and computation of $g = n/h$ pseudo-estimates; third, deletion of h observations in the $\binom{n}{n-h}$ possible different ways. The three schemes appear to be rather equivalent when the parameter h is moderate and the observations are strictly independent. When there is some serial correlation, the second scheme can be more reliable; the selection of h must be such that each group of h deleted observations be more or less independent of the other. The third scheme appears to be very valuable for theoretical derivations - see Sen (1977) - although it has limited practical interest. The above discussion is experimentally supported - see Miller (1974a) for references.

Before leaving this subject of bias reduction, it is worthy to mention generalizations of the jackknife method. Instead of evaluating the pseudo-estimates from estimators based on different sample sizes, it is possible to take into account different estimators on the same samples. This is defended at length in the collection of papers by Gray and Schucany as well as in their book (1972).

The second significant progress in the history of the jackknife takes place in 1958 when Tukey conjectures that the pseudo-values $\tilde{\theta}_i$ are essentially representative of the incidences of each specific observation, therefrom he proposes the jackknife variance estimate

$$\sigma^2(\hat{\theta}) = E[(\tilde{\theta}_i - \tilde{\theta})^2].$$

This conjecture will largely be supported in the sequel. We now simply retain the attention on the feature which is, in fact, the most significant property of the jackknife method.

Bias reduction as well as variance estimation can be achieved without detailed knowledge of the sample distribution nor involved analysis of the estimation method. We only need a sample and an estimation definition. - Next section will demonstrate that the third central moment of $\hat{\theta}$ is also available.

The computation involved by the jackknife method may be abusively large and, accordingly, one tends to apply it as musch as possible through analytical means. One interesting case is due to Dempster (1966). With difficulties to manipulate the deletion of one observation in his formulas, he states his developments with the help of a variable ε ranging from zero, no deletion, to one, complete deletion, and eventually limits himself to consideration of the first and second order terms in the variable ε. This approach is met anew with the infinitesimal jackknife of Jaeckel (1972a), where the above variable ε is kept infinitesimal. It is of great concern for analytical derivations because, then, the differences between estimators can be stated in terms of derivatives. But, due to the lack of theoretical background, it is scarcely published so far.

To conclude this introduction we would like to mention three papers which could introduce the reader to the jackknife. Mosteller (1971) in a naive presentation pleads for its introduction in elementary courses of statistics at the highschool level; Bissel and Ferguson (1975) place also some emphasis on the robust properties but say, after having demonstrated how beneficial the jackknife can be : "However the general warning still stands - the jackknife is sufficiently sharp to wound the unwary". The very important review of Miller (1974a) is possibly the most complete piece of information which tries to balance the advantages with the drawbacks of the method.

In the line of the jackknife but with very different objectives and theoretical set-up, we should mention the leave-one-out method of Lachenbruch (1968) which has only intuitive justification, the work in survey sampling of Woodruff and Causey (1976) which is adapted to the survey context and the interesting paper of Gray, Schucany and Watkins (1975). The last two papers are partly related to the infinitesimal jackknife.

3.2. Jackknife theory

Hereinafter, we sketch the main points of the derivation which justifies the jackknife method. A few points of this original work being rather involved, we refer to the appendix for further details. Particularly, this section will be solely concerned with scalar estimators although vector-valued estimators and functionals could be taken into account.

Assume we have at disposal a scalar estimator $\hat{\theta}$, consistent with respect to parameter θ and based on the sample $(x_1,\dots,x_n)$, each observation x_i appearing with the bounded non-negative weight w_i. Then the estimator may be seen as a transformation on the set of observations and weights. Let it be written as

$$\hat{\theta} = T(x_1,\dots,x_n;\ w_1,\dots,w_n).$$

Then, according to section 2.4, it can frequently be expanded in the form

$$\hat{\theta} = \theta + (1/\sum w_i)\sum w_i\ \psi_i + \frac{1}{2}\ (1/\sum\ w_i)^2\ \sum\sum\ w_i w_j\ \phi_{ij} + \dots$$

where the coefficients ψ_i and ϕ_{ij} are functions of the transformation $T(.)$ and the sample $(x_1,\dots,x_n)$, but independent of the weights.

This model, limited to its first few terms, will be assumed throughout. We do not know whether its validity is strictly required but we know it constitutes a sufficient condition for the validity of the jackknife method. Moreover, the jackknife fails in situations where the above model happens to be not verified.

Contrary to the derivation met in the introduction, we consider arbitrary modifications of the set of weights to derive the pseudo-values. In this section, however, we limit ourselves to the modification of one weight w_i, at a time (h=1, g=n). This modification can possibly correspond with the deletion of the observation. When the weight w_i is modified by a factor (1 + t), the pertaining pseudo-estimate is

$$\hat{\theta}_i = T[x_1,\ldots,x_n;\ w_1,\ldots,w_{i-1},\ (1+t)w_i,\ w_{i+1},\ldots,w_n].$$

We now define the pseudo-values by

$$\tilde{\theta}_i = [(tw_i + \sum w_j)\ \hat{\theta}_i - (\sum w_j)\ \hat{\theta}]/t$$

and the jackknife estimate through

$$\tilde{\theta} = (1/\sum w_i)\sum \tilde{\theta}_i.$$

The corresponding expansion has approximately the form

$$\tilde{\theta} = \theta + (1/\sum w_i)\sum w_i\psi_i + \frac{1}{2}(1+t)(1/\sum w_i)^2 \sum\sum w_i w_j \phi_{ij} + \ldots$$

which nearly equates the expansion met with for $\hat{\theta}$.

The first differing item is the second order term which appears multiplied by the factor (1 + t); this term is also the first which can introduce a bias decreasing with the sample size. Therefore its cancellation, obtained with t = -1, usually reduces the bias. We also note that small t does not bring any bias reduction.

With regard to bias reduction the following conclusions may be derived for a class of estimators

- If $\hat{\theta}$ can be expanded according to section 2.4,
- if the weights w_i are independent of the observations x_i,
- if the g groups have the same weights,
- if each group is independent of the others; then,

- the ordinary jackknife ($t = -1$) possibly reduces the bias,
- the jackknife estimate $\tilde{\theta}$ has the same asymptotic distribution as the original estimator $\hat{\theta}$, but for a translation resulting from bias reduction.

In order to obtain a variance estimate, we now consider more attentively the pseudovalues $\tilde{\theta}_i$. They have the particularly simple expansion

$$\tilde{\theta}_i = w_i \theta + w_i \psi_i + \dots ,$$

which indicates they are essentially proper to the perturbed observation weight w_i and independent of the other weights. Furthermore, the main random component of $\hat{\theta}$, i.e. $(1/\sum w_i) \sum w_i \psi_i$, is precisely the arithmetic mean of the main random components appearing in the pseudo-values. Thus, the second and third central moments of the former are, up to some factor, the homologous moments of the pseudo-values. This argument is the more correct, the more neglegible are the terms omitted in the expansions. Asymptotically, we have

$$\mu'_2(\hat{\theta}) = \mu'_2(\tilde{\theta}) = (1/\sum w_i)^2 \; g \; \mu'_2(\tilde{\theta}_i)$$

and

$$\mu'_3(\hat{\theta}) = \mu'_3(\tilde{\theta}) = (1/\sum w_i)^3 \; g \; \mu'_3(\tilde{\theta}_i).$$

The "not-quite large sample" situation is treated in the appendix and, in agreement with Tukey (1958), we obtain the jackknife variance estimate

$$\mathrm{var}(\hat{\theta}) \approx \mathrm{var}(\tilde{\theta}) = \sum(\tilde{\theta}_i - w_i\tilde{\theta})^2 / [(\sum w_i)^2 - \sum w_i^2].$$

Moreover the right hand expression is a fairly good approximation of $\mathrm{var}(\tilde{\theta})$, when $t = -1$. This is due to the cancellation of the second order random component in the jackknife estimate.

This variance estimator is obtained for the class of $\hat{\theta}$ which also tolerates bias reduction. Furthermore, its variability is low inasmuch as the number of groups g is sufficiently large and inasmuch as they are independent. Therefore, it may be safe to perturb simultaneously several observations when some correlation is expected. This consideration is opposite to Sharot's viewpoint (1976a), who only considers strict independency.

While computing the jackknife estimate, it is wise to check the range of the variates $(\tilde{\theta}_i - w_i \tilde{\theta})$ in order to assess the presence of possible outliers, they would impair the variance estimate. When robust estimators are jackknifed, these variates are bounded due to the limited incidence of each observation on the estimator $\hat{\theta}$.

For analytical derivations, it appears difficult to mimic the procedure as described above. The partial or complete deletion of an observation to obtain the pseudo-estimate $\hat{\theta}_i$ implies the use of a (finite) difference operator, whereas an (infinitesimal) differential operation would be markedly preferred. This is obtained through the use of a small perturbing parameter t.

In the infinitesimal version of the jackknife, t small, the expansions can strictly be limited to their first leading random terms as follows. The pseudo-estimates are

$$\begin{aligned}\hat{\theta}_i &= \hat{\theta} + tw_i(\partial/\partial w_i)\ T(x_1,\ldots,x_n;\ w_1,\ldots,w_n)\\ &= \hat{\theta} + tw_i\ j_i\end{aligned}$$

and the jackknife estimate is not bias reduced

$$\tilde{\theta} = \hat{\theta}.$$

The corresponding jackknife variance estimate is

$$\mathrm{var}(\hat{\theta}) = \sum(w_i j_i)^2/[1-\sum w_i^2/(\sum w_i)^2].$$

In his 1972 memorandum, Jaeckel has proposed to modify the jackknife procedure in order to obtain bias-reduction even with the infinitesimal version. His proposal involves the second partial derivatives of T with respect to w_i and w_j and is, in fact, as much as possible an analytical duplication of the computational treatment with (t = -1). We hesitate to recommend his derivation seeing that other appropriate treatments could as well be advanced to mimic the full deletion of an observation. - In a parallel line the paper of Sharot (1976b) is noteworthy, his comparison of several types of jackknife variance estimators leads him to conclude that :

> "On the basis of the Monte Carlo studies, it would appear that the desired gain in precision ... is often achieved. Alternative estimators designed for a particular application may, not surprisingly, do better still. The infinitesimal jackknife is seen to yield just such an estimator in many cases."

We frankly regret the infinitesimal jackknife is so little known. It only appears in one page of Miller's review (1974a), and there the reference to Jaeckel is of no great help.

To conclude this section on the jackknife theory, we indicate that Thorburn (1976) enumerates the conditions to be satisfied by the transformation T. He eventually obtains that the Taylor-like expansion we have assumed is required - see page 309, last but one equation. Nevertheless, we conjecture that the scope of application may be broader than implied by section 2.4.

3.3. Case study

To demonstrate the jackknife power, we investigate in this section an estimator of mean residual life frequently met with in survivorship studies. The particular interest of this application lies in the fact that the data may be censored in an unknown fashion. The derivation has already been reported in Rey (1975b) and another paper in the same field has been issued by Miller (1975); the latter has possibly been written in connection with the former.

Suppose we have a set of n independent items which are running from an initial time until they are either stopped or in failure. We will denote by x_i $(1 \leq i \leq m \leq n)$, the observed durations of the m failing items; the (n-m) remaining items are also observed during known durations x_i $(m < i \leq n)$. Then, under assumption of constant hazard rate, the mean residual time before failure is classically given by the maximum likelihood estimator

$$\hat{\theta} = (1/m) \sum x_i, \qquad i=1,\ldots,n.$$

note the sum runs on all items, whereas the denominator is the number of failing ones.

A question we will not resolve here is whether estimator $\hat{\theta}$ is consistent to the real mean life; this is not the object of the present discourse. This is important seeing that we now relax the assumption of constant hazard rate. We keep on the estimator definition but do not anymore assume that the failure times are distributed according to a negative exponential probability function. In this way, we expect more reliable results, derived solely from the data set, when the observations do not agree strictly with the exponential assumption. We free ourselves from distribution preconceived ideas. This is a feature of robustness.

Before applying the jackknife we note that the weighted form

$$\hat{\theta} = (1/\sum w_j) \sum w_i x_i, \qquad (j=1,\dots,m;\ i=1,\dots,n)$$

fits in the frame of section 2.4 and, thus, we are in a good position to proceed.

With all weights equal to one, we now develop the finite jackknife procedure. Modification of the weight w_i by a factor $(1 + t)$ produces the pseudo-estimate

$$\hat{\theta}_i = \hat{\theta} + t\,\delta_i$$

with

$$\begin{aligned} \delta_i &= (x_i - \hat{\theta})/(m + t), && \text{if } i \leq m, \\ &= x_i/m, && \text{if } i > m. \end{aligned}$$

The corresponding pseudo-value is

$$\begin{aligned} \tilde{\theta}_i &= [(n + t)\,\hat{\theta}_i - n\,\hat{\theta}]/t \\ &= \hat{\theta} + (n + t)\,\delta_i. \end{aligned}$$

Therefrom, we obtain the jackknife estimate

$$\begin{aligned} \tilde{\theta} &= \sum \tilde{\theta}_i/n \\ &= \hat{\theta} + [(n + t)/n] \sum \delta_i \\ &= \hat{\theta} + t\,[(n + t)/n] \sum x_i/[m(m + t)], \quad \text{for } i > m. \end{aligned}$$

The bias reduced estimate is the expression where $t = -1$ has been introduced, i.e.

$$\tilde{\theta}_0 = \hat{\theta} - \frac{n - 1}{n\,m\,(m - 1)} \sum x_i, \qquad \text{for } i > m.$$

We see that the bias correction is inversely proportional to the sample size, when the ratio of stopped to failing items is given. If this ratio is changing with size n, then the bias correction is meaningless. Effectively in such case the parameter θ to be estimated also fluctuates with size n and, therefore, what should be estimated is a matter of opinion.

A variance estimate is readily derived through the infinitesimal jackknife and this variance is correct to the first order in sample size for finite t-values. According to the definition

$$\hat{\theta}_i = \hat{\theta} + t w_i j_i,$$

we have in this case study

$$w_i j_i = \delta_i$$

and, immediately, the estimator follows

$$\mathrm{var}(\hat{\theta}) = \sum \delta_i^2/(1 - n/n^2)$$

or

$$\mathrm{var}(\hat{\theta}) = \frac{n}{m^2(n-1)} [\sum (x_j - \hat{\theta})^2 + \sum x_i^2], \quad (j \leq m,\ i > m).$$

The jackknife estimate and the jackknife variance estimate have been confirmed by Monte Carlo experiments. The quality of the results may be related to the fact that the δ_i-values are scarcely dependent upon t; this is indicative of rapid convergence of the expansion of section 2.4.

It is hardly possible to draw any conclusion from this simple illustration. However, we feel that the jackknife is worthy of consideration. We would like now to retain the attention more specifically on the covariance matrix which comes out for multidimensional estimators.

Assume we want to estimate the mean vector of a sample $(\underline{x}_1,\ldots,\underline{x}_n)$, as well as its covariance matrix, given that the multidimensional observations $\underline{x}_i$ are possibly incompletely known. We denote by $\hat{\underline{\theta}}$ the estimator of the mean and its j-th component will be given by

$$\hat{\theta}_j = \sum z_{ji} x_{ji} / \sum z_{ji}$$

to be unbiased. The coefficients z_{ji} are indices of presence or absence of component x_{ji}. Per definition, we have

$$z_{ji} = 1, \quad \text{if } x_{ji} \text{ is observed,}$$
$$\quad = 0, \quad \text{if } x_{ji} \text{ is missing.}$$

Before applying the jackknife method we note the need of independency assumptions on the z_{ji} as well as on the $\underline{x}_i$; furthermore the weighted form of $\hat{\theta}$,

$$\hat{\theta}_j = \sum w_i \; z_{ji} \; x_{ji} \; / \sum w_i \; z_{ji},$$

must be sufficiently differentiable with respect to the x_{ji} and to the w_i. These conditions are satisfied.

Attempts to reduce the bias would here put to light the lack of bias. We will therefore only estimate the covariance matrix in the sequel. It results from the partial derivatives.

$$J_{ji} = \partial \hat{\theta}_j / \partial w_i$$
$$= z_{ji} (x_{ji} - \hat{\theta}_j) / \sum w_i \; z_{ji}$$

and has the components

$$[\operatorname{cov}(\hat{\underline{\theta}})]_{kl} = \sum w_i^2 \; J_{ki} \; J_{li} \; / \; [1 - \sum w_i^2 \; / \; (\sum w_i)^2] .$$

We will write it down for the uniformly weighted estimator, that is to say for the unweighted mean vector,

$$[\operatorname{cov}(\hat{\underline{\theta}})]_{kl} = \frac{n}{n-1} \; \frac{\sum z_{ki} (x_{ki} - \hat{\theta}_k)(x_{li} - \hat{\theta}_l) \; z_{li}}{\sum z_{ki} \sum z_{li}} .$$

This covariance estimator is a positive definite matrix. It concurs with the ordinary estimator when all observations are complete. It does not seem useful to insist on the elegancy of the method seeing the interest of this estimator by comparison with the structure which can be inferred from the classical theory. The latter estimator is not even positive definite; it would have been written

$$[\operatorname{cov}(\hat{\underline{\theta}})]_{kl} = \frac{1}{m\;(m-1)} \sum_i z_{ki} \; (x_{ki} - \hat{\theta}_k) \; (x_{li} - \hat{\theta}_l) \; z_{li}$$

where

$$m = \sum_i z_{ki} \; z_{li} .$$

With regard to applications, they can be found in various domains;

some are interesting because it has been observed that the jackknife apparently fails on correlated data and on order statistics. This is not surprising to us. However it is informative to scan the papers of Ferguson et al. (1975), Gray et al. (1976), Miller (1974b), Rey (1974) and Rey and Martin (1975), posterior to the review of Miller (1974a).

3.4. Comments

On tendency to normality. It is known that under regularity conditions and with large group-size h as well as finite group-number g, the jackknife estimate tends to have a T-distribution with g-1 degrees of freedom. To us, this appears accidental and essentially a consequence of the performed mathematical derivation. Expressing the estimator $\hat{\theta}$ in terms of the observations x_i as a power series expansion and truncating at a low order produces necessarily pseudo-variates $\tilde{\theta}_i$ with normal distributions if their components are many and bounded (appropriately disguised in the regularity conditions). And then the conclusion that the jackknife estimate tends to normality is reached. This state of affair has been avoided in this paper by expanding in terms of the weights rather than in terms of the observations. It happens frequently that the jackknife estimate is more or less normal but, then, the original estimator $\hat{\theta}$ was also subject to application of the Lindeberg conditions.

On time-series analysis. Jackknifing can be applied under conditions of independency. When some stationarity in variances can be assumed and sample size permits, groups of relatively large sizes h must be assembled. All other conditions are generally fulfilled.

On order statistics. This presentations does not justify the utilization of the jackknife. Effectively, the estimators are usually not continuous in the observation weights and the weights are not independent of the observation values.

On misclassification probabilities. Numerous methods are in use to classify, among several classes, an extraneous sample on the basis of prior information. This prior information consists frequently in a set of observations belonging to known classes. Then, the probability of misclassification can be seen as a function of these known observations. Estimator of the misclassification probability can be jackknifed in "discriminant analysis", where the estimator depends smoothly on observation importances; but application of the method with some other classification techniques, such as "nearest neighbour", is not justified.

On transformations. If the jackknife method is applicable to $\hat{\theta}$, it is also applicable to $\hat{\phi} = \phi(\hat{\theta})$ inasmuch as $\phi(.)$ is continuously once differentiable in the vicinity of θ. However, the transformation may be useful to set confidence domains when the distribution of $\hat{\phi}$ is more easy to manipulate than the original distribution.

On robustness. The pseudo-values $\tilde{\theta}_i$ are indicative of the observation incidences on the estimator $\hat{\theta}$ and happen to be a discrete version of Hampel's influence curve (1974). Their inspection may reveal an abusively high sensitivity to some observations.

4. M-ESTIMATORS

4.1. Warning

In these sections we meet with a few classical problems which are well-known in the theory of statistical estimation. However the viewpoint is rather unusual and this leads to disregard certain aspects which otherwise would be of great concern.

For example, consider the estimation of a location parameter θ for the variate x distributed according to the law $f(x - \theta)$. We will search for the "best" estimator without feeling concerned by its representativity or its real meaning. We may end up with a median estimate or an arithmetic mean; both are invariant with respect to the real location under translation of the distribution; both may frequently be used to estimate the location, however they may differ from each other.

But what are these M-estimators we investigate ? They are generalizations of the usual maximum likelihood estimates. Classically θ is the parameter value maximizing the likelihood function, i.e. we have in obvious notation

$$L = \Pi \; f(x_i \mid \theta) = \max \text{ for } \theta$$

or equivalently

$$- \ln L = - \sum \ln f(x_i \mid \theta) = \min \text{ for } \theta.$$

The estimators of type M are solutions of the more general structure

$$M = \sum \rho(x_i, \theta) = \min \text{ for } \theta,$$

where the function $\rho(.)$ may be rather arbitrary. Before proceeding, we note that the above structure is rarely appropriate to process correlated observations.

The estimators of type M have been analysed in quite many respects for location problems, since the initial contribution of Huber (1964); however, the many other estimation problems have received scarcely any attention - The next section is original, in this respect - We will meet with developments which are based on differentiability and independency properties which usually do not hold for the other two great classes, the L- and the R-estimators. The latter are respectively obtained through linear combinations of order statistics and through rank tests. We refer to Huber (1972) for their properties (some are questionable), and to Scholz (1974) for their respective merits.

Developping the theory of M-estimators we will need various functions, their derivatives, operators with vector or matricial structures as well as summations and integrals. In order to ease the reading we present the notation in use. As much as possible we have tried to maintain the now classical notation; further, each time we need a special script, we have recalled its definition.

We will successively meet with the following scripts

Ω	sample space, $\Omega \subset R^p$.
p	dimensionality of Ω.
$f(x)$	probability density function of x, $x \in \Omega$.
x_i	observations, $1 \leq i \leq n$.
n	sample size.
w_i	non-negative weight of observation x_i.
$\delta(x - x_i)$	Dirac function concentrated on x_i.
θ_j	parameter and estimator, $1 \leq j \leq g$.
g	number of simultaneous estimators.
M_j	function minimized by θ_j, $1 \leq j \leq g$.
$\rho_j(x,.)$	contribution of x to M_j, $1 \leq j \leq g$.
$\psi_j(x,.)$	$= (\partial/\partial\theta_j)\ \rho_j(x,.)$.
$\phi_{jk}(x,.)$	$= (\partial/\partial\theta_k)\ \psi_j(x,.)$.
$\Omega_j(x)$	Hampel's influence function, $x \in \Omega$.
t	$t \in R$, $0 \leq t \leq 1$.
$(.)^*$	perturbed entities have a star superscript.
$(.)'$	transposed entities have a prime superscript.
$\underline{(.)}$	underlined entities are column vectors.
σ_j^2	asymptotic variance of θ_j.
$u, \underline{v}, \varepsilon$	regression model, $u = \underline{v}'\ \underline{\theta}_1 + \varepsilon$.
m	dimensionality of $\underline{v}$ and $\underline{\theta}_1$, $p = m + 1$.
$\underline{x}$	$= (u, \underline{v}')' \in \Omega$.
$\psi(\varepsilon)$	$= (\partial/\partial\varepsilon)\ \rho_1(\varepsilon)$.
$\phi(\varepsilon)$	$= (\partial/\partial\varepsilon)\ \psi(\varepsilon)$.
R_i	rigidity index, $R_i = (\partial/\partial u_i)\ \varepsilon_i$.
ν	exponent in $\rho_1(\varepsilon) = k\ \lvert\varepsilon\rvert^{\nu}$.

θ	$= \theta_1$ for $m = 1$, $v = 1$.
$\underline{y}, D, F, \underline{e}_i, n$	see section 4.3.4.
V	$= \sigma_1^2$, asymptotic variance of θ.
s	scale of ε.

4.2. MM-estimators

Consider some sample space, say $\Omega \subset R^p$, on which a density distribution $f(x)$, $x \in \Omega$, is defined; possibly this distribution is unknown except for some sample $(x_1,\ldots,x_n)$ and then we take into account the empirical distribution.

$$f(x) = (1/\sum v_i) \sum v_i \; \delta(x - x_i)$$

based on a set of non-negative weights $(w_1,\ldots,w_n)$ and where $\delta(x - x_i)$ is a Dirac function concentrated on the observation x_i. In this section we investigate the estimation of a set of parameters $(\theta_1,\ldots,\theta_g)$ which are such that they minimize the functions $(M_1,\ldots,M_g)$, that is they are such that

$$M_j = \min \text{ for } \theta_j,\; j = 1,\ldots,g$$

where

$$M_j = \int \rho_j \;(x,\, \theta_1,\ldots,\theta_g)\; f(x)\; dx.$$

This definition yields to identity of the parameters with their estimates.

The above framework is essentially a generalization of the now classical M-estimator theory. It is motivated by the frequent interdependence of various estimators. For instance, a scale estimator often depends upon a previous estimation of some location estimator, e.g. the variance σ^2 may be defined through the M-structure, for $p = 1$,

$$\int [(x - \mu)^2 - \sigma^2]^2 \; f(x)\; dx = \min \text{ for } \sigma^2,$$

where μ is the mean through another M-structure

$$\int (x - \mu)^2 \; f(x)\; dx = \min \text{ for } \mu.$$

These are in fact estimators of type Multiple-M or, as we call them, MM-estimators.

In the sequel we assume that the functions M_j can be differentiated under the integral sign. More precisely, we assume :

- Independence of Ω with respect to $\theta_1,\ldots,\theta_g$, (conditions on derivatives of f(x) at the frontier of the sample space Ω generally meet the needs).
- $\psi_j(x,.) = (\partial/\partial\theta_j)\ \rho_j(x,.)$,
  $\phi_{jk}(x,.) = (\partial/\partial\theta_k)\ \psi_j(x,.)$,
  (existence and differentiability of derivatives).

Accordingly, the MM-estimators can as well be defined by the following set of g equations

$$\int \psi_j(x,\ \theta_1,\ldots,\theta_g)\ f(x)\ dx = 0.$$

In our illustration, the corresponding set has g = 2 equations and is, for an empirical distribution f(x),

$$\sum w_i\ [(x_i - \mu)^2 - \sigma^2] = 0,$$

$$\sum w_i\ (x_i - \mu) = 0.$$

Strictly speaking, μ is an M-estimator, whereas σ^2 is an MM-estimator.

We will now be interested by the sample distributions of these estimators. Precisely, we first derive their influence functions, then we give their variances and eventually conclude this section by some considerations on their robustness.

The notation is relatively difficult to select without restricting ourselves to certain domains of applications. In order to permit easy adaptations of the derivations, we will assume that the estimators θ_j are column-vectors (possibly of dimension 1). The transposition will be denoted by a prime superscript.

According to section 2.5, the Hampel's influence function relative to the estimate θ_j is given by

$$\Omega_j(x_0) = \lim[(\theta_j^* - \theta_j)/t],\ t \to 0$$

where θ_j^* is defined through the perturbed distribution

$$f^*(x) = (1 - t)\ f(x) + t\ \delta(x - x_0)$$

for a given coordinate set x_0. Observe that the script $\Omega_j(x)$ describes a vector-valued function on the sample space and should not be confused with the sample space itself; the notation $\Omega(.)$ is becoming standard, although unfortunate; it strictly corresponds with the cumbersome $IC_{\theta_j,F}(.)$ of Hampel (1974).

The g expressions

$$\int \psi_j(x, \theta_1^*, \ldots, \theta_g^*)\ f^*(x)\ dx = 0$$

can be expanded as follows with respect to the non-perturbed elements - We contract the notation and only take into account the terms up to the first order in the perturbation.

$$\begin{aligned} \int \psi_j^*(x)\ f^*\ dx & \\ &= \int [\psi_j(x) + \sum \phi_{jk}(x)\ (\theta_k^* - \theta_k)]\ [(1 - t)\ f + t\ \delta(x_0)]\ dx \\ &= \sum [\int \phi_{jk}(x)\ f\ dx]\ (\theta_k^* - \theta_k) + t\ \psi_j(x_0) \\ &= \sum A_{jk}(\theta_k^* - \theta_k) + t\ \psi_j(x_0) = 0. \end{aligned}$$

We see that the difference $(\theta_j^* - \theta_j)$ is the solution of a set of g linear equations. They may be scalar but possibly they are vectorial or functional depending upon the nature of the estimators. When the coefficients A_{jk} $(k \neq j)$ are dominated by A_{jj}, that is when the estimators are relatively independent from one another, a solution can be derived. Under

$$\| \textstyle\sum_k A_{ik}\ A_{kk}^{-1}\ A_{kj}\ A_{jj}^{-1} \| \ll 1,\ k\neq i,\quad k\neq j,$$

we obtain

$$\theta_j^* - \theta_j = -\ t[A_{jj}^{-1}\ \psi_j(x_0) - \textstyle\sum_k A_{jj}^{-1}\ A_{jk}\ A_{kk}^{-1}\ \psi_k(x_0)],\quad k\neq j.$$

Therefrom, the influence function is given by

$$\Omega_j(x_0) = \sum A_{jj}^{-1}\ A_{jk}\ A_{kk}^{-1}\ \psi_k(x_0) - 2\ A_{jj}^{-1}\ \psi_j(x_0)$$

where the summation runs over k = 1,...,g and

$$A_{jk} = \int \phi_{jk}\ (x, \theta_1, \ldots, \theta_g)\ f(x)\ dx.$$

When the distribution f(x) is experimental, the integrals equate the corresponding sums, e.g.,

$$A_{jk} = (1/\sum w_i) \sum w_i \, \phi_{jk}(x_i, \theta_1, \ldots, \theta_g),$$

and the perturbation of f(x) in $f^*(x)$ can be limited to a modification of the weight set $(w_1, \ldots, w_n)$. This is precisely what has been the treatment described in section 3.2 on the jackknife. Accordingly we meet here with the variance estimate

$$\mathrm{var}(\theta_j) = \sum w_i^2 \, [\Omega_j(x_i)] \, [\Omega_j(x_i)]' \; / \; [(\sum w_i)^2 - \sum w_i^2].$$

The notation with a prime, [.]', is for the transposed of the influence function when the latter is not a scalar.

If and only if all observations have the same weight, the variance estimate is related to the asymptotic variance

$$\sigma_j^2 = \int \Omega_j(x) \, \Omega_j(x)' \, f(x) \, dx$$

through the asymptotic expression

$$\mathrm{var}\,(\theta_j) = \frac{1}{n-1} \, \sigma_j^2 .$$

So far we have not devoted much attention to the nature of the solution. A natural requirement, apart from being robust, is to be unique or at least "locally" unique; that is we require the set of solutions to be discrete. A locally convex set of solutions may be accepted in certain contexts but this is rather troublesome and, often, no loss of generality happens when the indeterminacy is resolved by a supplementary condition. In case the indeterminacy must be preserved, it is appropriate to work out the solutions with the help of their projections on some subspace and, there, no difficulty occurs.

Given that the solution is a minimum, and with the requirement of local unicity, we see that the coefficients A_{jj} must be positive definite, thus they can be inverted as was implicitely assumed in the derivation.

To be robust we must also have a bounded influence function and, therefore, the first derivatives $\psi_j(x,.)$ must be bounded for any x in the sample space. Then, the MM-estimator θ_j has a finite variance and its distribution tends to the normal law (possibly multivariate) according to the Lindeberg condition (eq. 6.3 of Feller, 1966) and the multivariate central limit theorem (Rao, 1973, p. 128).

An apparent conceptual difficulty should be clarified concerning the estimator definition. Having at disposal a distribution or a sample, f(x), and a mathematical rule, the minimisation of $(M_1, \ldots, M_g)$, we have defined the parameter to be estimated and its estimator as corresponding to the minimum and, thus, equivalent. Further, in spite of this "confusion", we have designed an estimator of some variance. What is the argument behind this mess ?

The viewpoint is here, from beginning to end of the argument, that the unique information we have on the distribution is f(x), say for a sample

$$f(x) = (1/\sum w_i) \sum w_i \; \delta(x - x_i),$$

and that it appears fictitious to introduce any parent distribution, say g(x), if it is not peremptory. This is opposite to the usual sampling theory and, thus, startling. The definition of the parameter to be estimated is given by reference to f(x) rather than to g(x); this avoids a parent distribution with unclear relations to f(x), or with arbitrarily cleared relations. But we implicitely assume existence of this parent distribution g(x), further it must fit in the frame of section 2.4. Then, with unique reference to f(x), it is possible to assess the variability of the estimators accordingly with section 3.2. In fact, contrary to the usual sampling theory, we do not need here any parent distribution in an explicit form and, thus, it has been omitted.

Before proceeding, we propose a simple illustration where all the concerned elements can be explicitely stated, With the above formalism, we study the central moment of order ν, in the ordinary notation μ_ν. Its definition

$$\mu_\nu = \int (x - \mu)^\nu \; f(x) \; dx$$

implies the knowledge of the mean

$$\mu = \int x \; f(x) \; dx.$$

We denote these parameters respectively θ_1 and θ_2 and transform the definitions as follows

$$\int [\theta_1 - (x - \theta_2)^\nu] \; f(x) \; dx = 0,$$

and

$$\int [\theta_2 - x] \; f(x) \; dx = 0.$$

Comparing these last two expressions with the fundamental MM-estimator equation

$$\int \Psi_j(x, \theta_1, \ldots, \theta_g) \; f(x) \; dx = 0,$$

we observe the correspondence and we define

$$\Psi_1(x, \theta_1, \theta_2) = \theta_1 - (x - \theta_2)^\nu,$$

and

$$\Psi_2(x, \theta_1, \theta_2) = \theta_2 - x.$$

These equations have the following partial derivatives

$$\psi_{11}(x,.) = 1, \qquad \psi_{12}(x,.) = \nu(x - \theta_2)^{\nu-1},$$

$$\psi_{21}(x,.) = 0, \qquad \psi_{22}(x,.) = 1;$$

which will be introduced in

$$A_{jk} = \int \psi_{jk}(x,.) \; f(x) \; dx$$

to obtain the factors

$$A_{11} = 1, \qquad A_{12} = \nu \; \mu_{\nu-1},$$

$$A_{21} = 0, \qquad A_{22} = 1.$$

This material leads us to the expression of the influence function

$$\Omega_1(x) = A_{11}^{-1} \; A_{12} \; A_{22}^{-1} \; \Psi_2(x) - A_{11}^{-1} \; \Psi_1(x)$$

$$= (x - \mu)^\nu - \mu_\nu - \nu \; \mu_{\nu-1}(x - \mu)$$

which indicates the incidence of an observation x on the central moment μ_ν. The asymptotic variance

$$\sigma_1^2 = \int \Omega_1^2(x) \; f(x) \; dx$$

does not present any computational difficulty. Now we more specifically devote our attention on an estimation of μ_ν based on a sample of size n of equally weighted observations. We denote m_ν the estimate of μ_ν and the distribution is the probability density function

$$f(x) = (1/n) \sum \delta(x - x_i).$$

The variance of m_ν comes out immediately, without any explicit reference to a parent distribution

$$\mathrm{var}(m_\nu) = \frac{1}{n-1} \int \Omega_1^2(x)\ f(x)\ dx$$

$$= \frac{1}{n-1} (m_{2\nu} - m_\nu^2 + \nu^2\ m_2\ m_{\nu-1}^2 - 2\nu m_{\nu-1}\ m_{\nu+1}).$$

This form concurs with the classical result, up to the second order in n.

4.3. M-estimators in location and regression

In this section we specialize the above derivation to the situation g = 1 and obtain the conditions to be satisfied by $\rho_1(.)$, or $\psi_1(.)$, in order to have robust estimates. This presentation provides means of assessing the robustness but, moreover, it leads to the design of the functions $\rho_1(.)$ and $\psi_1(.)$ which produce the "best" estimators, best in some sense.

4.3.1. Location and regression

Let us first set the stage. We are concerned by the linear model

$$u = \underline{v}'\ \underline{\theta}_1 + \varepsilon$$

where u and ε are scalars and $\underline{v}$ and $\underline{\theta}_1$ are m-dimension column-vectors. In case of a location problem, we simply have m = 1 and v = 1. Each observation consists in the set of parameters

$$\underline{x}_1 = (u_i, \underline{v}'_i)' \in \Omega$$

and, per definition, $\underline{\theta}_1$ is the set of values minimizing

$$M_1 = \int \rho_1(\underline{x}, \underline{\theta}_1)\ f(\underline{x})\ dx$$

or

$$M_1 = (1/\sum w_i) \sum w_i\ \rho_1(\underline{x}_i, \underline{\theta}_i)$$

with

$$\rho_1(\underline{x}, \underline{\theta}_1) = \rho_1(\varepsilon) = \rho_1(u - \underline{v}' \underline{\theta}_1).$$

The fact that u and $\underline{v}$ are assembled in the script $\underline{x}$ is more than a writing commodity. Both are concerned at the same level and an observation $\underline{x}_i$ may be outlying as well because $\underline{v}$ is abnormal as because u is; one should correspond with the other. The situation of outlying $\underline{v}$ has been reviewed by Hill (1977) as well as Ypelaar and Velleman (1977).

As previously, we observe that $\underline{\theta}_1$ is defined through the mathematical rule of minimization rather than through a statistical characterization of the possibly-random variable ε. Inasmuch as $\rho_1(.)$ is an increasing function of $|\varepsilon|$, the minimization process can also be seen as a way of approximating u by $\underline{v}' \underline{\theta}_1$, in spite of possible inadequacy of the linear model.

In location, the emphasis on the mathematical rule rather than on the statistical characterization has been skipped by Huber (1964), and partly by Jaeckel (1971), by considering symmetric distribution f(x) where there is a "natural" definition of $\underline{\theta}_1$, the center of symmetry. Then limiting $\rho(.)$ to be symmetric clears most statistical problems. This emphasis is underlying the discussion of what is robustness (or what it should be) by Huber (1972, 1973, 1977a). The minimization viewpoint is more critically analysed with the more recent investigations of regression problems; Jaeckel (1972b) and Collins (1976) state their concerns. But rather generally, authors ignore the possibility of a statistical characterization when they study finite samples and only take in consideration asymptotic properties, e.g. Maronna (1976). In most cases the finite sample properties are conjectured from Monte Carlo simulations. A nice case in the literature is the Hampel's papers (1973, 1975) pleading for the use of appropriate mathematical rules and which have been directly opposed by Dempster (1975, 1977) who only works on the statistical characterization in a bayesian framework. Obviously if one admits to select a data-dependent prior, very good and robust estimators can be derived. The dependence on data is often introduced in a sequential way (the prior is modified until obtention of "satisfactory" results) and frequently consists in arbitrary trimming or winsorization, see Yale and Forsythe (1976). This may require specific identification of outlying observations and many methods have been investigated in this regard since Anscombe (1960); let us just mention the work of Garel (1976). However it seems to us a need to endure values, as does Youden (1972), in keeping a critical eye on what is produced, as recommend Mead and Pike (1975).

After having travelled a long way between the regression seen either as a statistical procedure or seen as a type of approximation, let us resume our development of this particular M-estimator. In this derivation we will take into account samples of moderate sizes. For asymptotic properties we refer the reader to Huber (1964, 1972) and, more specifically for regression problems, to Huber (1973). We now move along section 4.2.

Noting $\psi(\varepsilon)$ and $\phi(\varepsilon)$ for the first and second derivatives of $\rho_1(\varepsilon)$ with respect to ε, the derivatives with respect to $\underline{\theta}_1$ have the simple expressions

$$\psi_1(\underline{x}, \underline{\theta}_1) = (\partial/\partial\underline{\theta}_1)\ \rho_1(u - \underline{v}'\ \underline{\theta}_1)$$
$$= -\ \psi(u - \underline{v}'\ \underline{\theta}_1)\ \underline{v}$$

as well as

$$\phi_{11}(\underline{x}, \underline{\theta}_1) = (\partial/\partial\underline{\theta}_1)\ \psi_1(\underline{x}, \underline{\theta}_1)$$
$$= \phi(u - \underline{v}'\ \underline{\theta}_1)\ \underline{v}\ \underline{v}'.$$

The factor A_{11} is a generalization of the ordinary design matrix. It has the form

$$A_{11} = (1/\sum w_i)\sum w_i\ \phi_{11}(\underline{x}_i, \underline{\theta}_1)$$
$$= (1/\sum w_i)\sum w_i\ \phi(\varepsilon_i)\ \underline{v}_i\ \underline{v}'_i$$

which must be positive definite. The influence function, at the observation $\underline{x}_i$, has the form

$$\Omega_1(\underline{x}_i) = \psi(\varepsilon_i)\ A_{11}^{-1}\ \underline{v}_i$$

and leads to a covariance estimate, here written for equal weights,

$$\mathrm{var}(\underline{\theta}_1) = \frac{n}{n-1}\sum \psi(\varepsilon_i)^2\ [\sum \phi(\varepsilon_j)\ \underline{v}_j\ \underline{v}'_j]^{-1}\ \underline{v}_i\ \underline{v}'_i\ [\sum \phi(\varepsilon_j)\ \underline{v}_j\ \underline{v}_j]^{-1}$$

which is valid for independent observations $\underline{x}_i$.

Robust regressions is a field where the concern is more on getting a regression certainly right in "the bulk of the data" than taking the risk of being misled by some odd observation. Effectively, it occurs that the classical methods of least squares, of maximum likelihood and of other probabilistic origins are very sensitive to outlying observations by attributing them a disproportionate importance. The two most obvious aspects are, on the one hand, that outlying observation $\underline{x}_k$ may contribute by a very important term $\rho_1(\varepsilon_k)$ or, on the other hand, that its residual ε_k may be abnormally small. The latter is particularly difficult to detect, although it is the most important : it means that the regression has been fitted on the outlier instead of neglecting it. This conflict between the desire of obtaining a not-small ε_k (without having identified $\underline{x}_k$ as outlier) and the desire of producing small ε_i through minimization is at the root of all robust regression methods.

Hopefully, an appropriate choice of the function $\rho_1(.)$ should resolve the above conflict. We already have at disposal the influence function to evaluate the dependency of $\underline{\theta}_1$ on any specific observation $\underline{x}_i$, however it usually tends to cancel for small residuals and, thus, cannot be used to recognize that the regression has been "locked" on some $\underline{x}_k$. But other ways can be proposed to assess this dependency; we have particularly in view a "rigidity index" which reflects the modification of ε_i resulting from a change of u_i. Specifically, it is defined by

$$R_i = \partial \varepsilon_i \;/\; \partial u_i$$

and is the closer to one, the more "rigid" the regression is with respect to erroneous u_i. Conversely, while the fit is locked on an outlier, the above index is close to zero. In the present framework, this index can be evaluated by

$$R_i = 1 - w_i \; \phi(\varepsilon_i) \; \underline{v}'_i \; A_{11}^{-1} \; \underline{v}_i$$

as may be seen by differentiation of

$$\sum w_i \; \psi_1(\underline{x}, \theta_1) = 0$$

or of

$$\sum w_i \; \psi(\varepsilon_i) \; \underline{v}_i = 0.$$

We may now infer what conditions the function $\rho(\varepsilon)$ must satisfy in order to produce a robust estimation, robust in the sense that each observation $\underline{x}_i$ has a bounded incidence on the estimator whatever it is and whatever is its non-negative weight w_i.

Bounding the rigidity index to satisfy

$$0 < R_i \leqslant 1$$

implies bounding of ρ-second derivative, i.e.

$$\phi(\varepsilon) \geqslant 0$$
$$\phi(\varepsilon) \text{ bounded, a.e.}$$

In fact $\phi(\varepsilon_i)$ must be strictly positive for at least m independent observations $\underline{x}_i$ in order to have a positive-definite matrix A_{11}. Bounding the influence $\Omega_1(\underline{x}_i)$ implies bounding of ρ-first derivative, i.e.

$$\psi(\varepsilon) \text{ bounded, a.e.}$$

In order to obtain an unbiased estimator, when the linear relation

$$u = \underline{v}' \; \underline{\theta}_1$$

is possible (with zero residuals), we also impose

$$\psi(0) = 0.$$

The fact that the bounding must be achieved for "any possible" ε is reminded by the almost everywhere (a.e.) indication. It does not matter if the derivatives are not bounded, or undefined, for some ε realized with probability zero, such as ε corresponding with $\underline{x} \notin \Omega$. A function $\rho_1(\varepsilon)$ satisfying this set of conditions will be said "admissible".

To conclude this discussion, we sum up : *To be admissible, the function $\rho_1(\varepsilon)$ must be piecewise convex (i.e., not concave), be minimum for $\varepsilon = 0$, have almost everywhere bounded first two derivatives and be strictly convex in the vicinity of at least m residuals ε_i.* We further add to guarantee $\underline{\theta}_1$ unicity : *the function $\rho_1(\varepsilon)$ must be continuous and convex everywhere.*

A natural question is whether there exists any admissible $\rho_1(\varepsilon)$. The answer is positive, but admissible $\rho_1(\varepsilon)$ cannot produce scale invariant $\underline{\theta}_1$ when the sample space is not bounded, i.e. with

$\Omega = R^p$, $p = m + 1$. Effectively, assume we require $\underline{\theta}_1$ to be independent with respect to the scale of the residuals, that is to be independent of the non-zero real variable λ in

$$\sum w_i \ \psi(\lambda\varepsilon_i) \ \underline{v}_i = 0;$$

then we must have cancellation of the first derivative

$$\partial\underline{\theta}_1 \ / \ \partial\lambda = 0$$

or

$$\sum w_i \ \varepsilon_i \ \phi(\lambda\varepsilon_i) \ \underline{v}_i = 0.$$

We now compare the last and the last but two expressions. For an arbitrary set of observation weights, their compatibility implies that $\varepsilon_i\phi(\lambda\varepsilon_i)$ is proportional to $\psi(\lambda\varepsilon_i)$. Therefrom we obtain that the only scale invariant structure, for $\underline{x} \in R^p$, satisfies

$$\varepsilon \ \phi(\varepsilon) \ / \ \psi(\varepsilon) = \text{constant}$$

or

$$\varepsilon \ \phi(\varepsilon) = (\nu-1) \ \psi(\varepsilon)$$

and is

$$\rho_1(\varepsilon) = k|\varepsilon|^\nu$$

with

$$k \neq 0, \quad \nu \neq 1.$$

But this specific form is not admissible, seeing that the derivatives can be arbitrarily large. We thus face an alternative : either we must bound the sample space Ω, or we must use a scale-dependent function $\rho_1(\varepsilon)$ in order to produce a robust estimation. The first term of the alternative will now retain our attention; the second will be investigated in section 4.4 where the scale invariant structure $\rho_1(\varepsilon/s)$ is analysed and which implies the knowledge of s, the "scale" of the variable ε.

4.3.2. Least powers.

In this section we further investigate the estimation of $\underline{\theta}_1$ minimizing

$$M_1 = \sum_i w_i \ \rho_1(\varepsilon_i)$$

with

$$\rho_1(\varepsilon) = |\varepsilon|^{\nu}$$

and

$$\varepsilon = u - \underline{v}' \, \underline{\theta}_1 .$$

We assume that the parameter ν satisfies

$$\nu \geqslant 1$$

in order to have a non-concave $\rho(\varepsilon)$ and that some bounding of the sample space avoids exceedingly large residuals ε_i.

In this family of $\rho(\varepsilon)$ are found several well known schemes. The most famous method is certainly the least squares minimization with $\nu = 2$, but the Chebyshev's criterion and the absolute value minimization, respectively ν tending to infinity and to 1, have equally received much attention.

Large values of the parameter ν will not be considered in the sequel due to the fact they do not present any interest in common statistical applications. In fact the smaller ν is, the smaller is the incidence of the large residuals on the $\underline{\theta}_1$ estimate; it appears that ν must be fairly moderate to provide a relatively robust estimator or, in other terms, to provide an estimator scarcely perturbed by outlying observations. The selection of an optimal ν is investigated by Ronner (1977).

The value $\nu = 2$, corresponding with the least squares or the arithmetic mean, is still too large as we can appreciate by the opinion of the Princeton group based on Monte Carlo runs. In Andrews et al. (1972, p. 239), they answer the question

> Which was the worst estimator in the study ?
> If there is any clear candidate for such an overall statement, it is the arithmetic mean, long celebrated because of its many "optimality properties" and its revered use in applications. There is no contradiction : the optimality questions of mathematical theory are important for that theory and very useful, as orientation points, for applicable procedures. If taken as anything more than that, they are completely irrelevant and misleading for the broad requirements of practice. Good applied statisticians will either look at the data and set aside (reject) any clear outliers before using the "mean" (which, as the study shows, will prevent the worst), or

they will switch to taking the median if the distribution looks heavy-tailed."

The difficulties have been surveyed by Huber (1972) and it may be concluded that only limited importance must be attributed to the most extreme data. For ν in the vicinity of 1.2, a good estimate may be expected, but this is a matter of opinion

Technically the only parameter which bounds the large residual incidence is $\nu = 1$. But then strict convexity is lost and an indeterminate solution may result. The corresponding problems are well known and are usually solved by linear programming techniques. However the indeterminacy may also be resolved in considering $\nu = 1$ as the inferior limit of $\nu > 1$. Inasmuch $\underline{\theta}_1$ does not vary to much with ν, this procedure appears reasonable; it has been recommended by Jackson (1921) who investigated the one-dimension median. - Our own experience totally supports this view - An interesting aspect is that this provides a natural extension of the one-dimension median to the multivariate domain. The dependency of $\underline{\theta}_1$ on ν has been the object of great attention recently with the present existence of computational facilities. Nowadays, we can compute, but is it meaningful in theory ? The answer is positive but reluctant for most authors, a reluctancy related to the pertinence of the needed theoretical assumptions. Fletcher et al. (1974) approach the question with a computation oriented viewpoint, while Cargo and Shisha (1975), Hwang (1975) as well as Lewis and Shisha (1975) analyse topological aspects. The theoretical considerations have a serious impact on practice due to the fact that a great number of tricks are used in the algorithms to force their eventual convergences to possibly artificial solutions. Before leaving these considerations, note that instead of minimizing the sum of powers of residuals an immediate generalization permits to minimize vector norms and thus to extend $\underline{\theta}_1$ from a vector to a matrix structure. These norm minimizations are investigated by Boyd (1974) and Rey (1975a); the last paper is partly oriented toward multidimensional location problems.

As we said, many difficulties are encountered in the computations when the parameter ν is in the interesting range $1 < \nu < 2$, for zero residuals are troublesome.

Several well known methods are at disposal to minimize functions. Let us first discard all the methods relying on some separation of function components (e.g., orthogonal decomposition); due to the fairly involved non-linearities, it does not appear possible to us to proceed in this way for the parameter ν in the interesting range. The search techniques of minimization must also be rejected because they have a very poor accuracy and are inclined to numerical trouble for a "flat" function minimum, situation which occurs quite often. Then we are left with relaxation and gradient methods. The former, such as used by Gentleman (1965), are equivalent to first order gradient methods in this context and will be presented at section 4.4.1 for MM-estimators.

Unfortunately, the second order gradient algorithms cannot be considered due to the problems arising in the determination of the second order derivatives. They involve the absolute residuals, $|\varepsilon_i|$, to a negative power and they are responsible for hazards in the algorithm convergence whenever the solution $\underline{\theta}_1$ correspond to small residuals. To bypass this difficulty, Forsythe (1972) recommends the implementation of the Davidon method (Fletcher and Powell, 1963), whereas Ekblom (1973, 1974) proposes the elimination of the poles by a quadratically perturbed method. It is noteworthy to observe that the Ekblom method corresponds to a somewhat restricted case of the generalized problem of Rey (1975a). Indeed, it consists in addition of a fictitious second dimension to the scalar data, u_i. We have prefered to apply the first order gradient method with a specially chosen step size, that is, chosen in order to "avoid" the poles of the second derivatives. However, with experience progressing, we have noted that the algorithm to solve the highly non-linear equations met with at section 4.5 was the most efficient in computation time.

With regard to computational experience, Merle and Späth (1974) present a clear discussion of several methods to be found in the prior literature. Comparison of the least powers approach with other techniques has been proposed in Rey (1977).

4.3.3. Can we expand in Taylor series ?

Whether the least power estimator $\underline{\theta}_1$ fits in the frame of section 2.4 may be inferred from the theoretical papers we have already mentioned but, seeing the importance of the question, we believe useful to directly demonstrate under what conditions it does fit. We will investigate the situation of the minimum sum of absolute residuals, $\nu = 1$. For larger power parameter, $\nu > 1$, the continuity of $\underline{\theta}_1$ with respect to ν is sufficient to demonstrate the Taylor-like expansion given that it is valid for least squares, $\nu = 2$.

In the sequel, we implicitely assume that, in situation of indeterminacy, the entities for $\nu = 1$ equate per definition the limit obtained for ν tending to 1 from above.

Let $\underline{\theta}_1$ be the solution of

$$\int \underline{v}\, \psi(u - \underline{v}'\, \underline{\theta}_1)\, f(\underline{x})\, d\underline{x} = 0$$

where, as previously, the integral spans the sample space Ω with

$$\underline{x} = (u, \underline{v}')' \in \Omega.$$

The probability density function $f(\underline{x})$ is assumed sufficiently regular in the domain satisfying

$$|u - \underline{v}'\, \underline{\theta}_1| \text{ small.}$$

When $\nu = 1$, the function $\psi(\varepsilon)$ becomes undefined at $\varepsilon = 0$. Any finite value may be used and this will not matter as long as it is finite. For instance, the above limiting scheme leads to

$$\begin{aligned} \psi(u - \underline{v}'\, \underline{\theta}_1) &= -1, \text{ if } u < \underline{v}'\, \underline{\theta}_1, \\ &= 0, \text{ if } u = \underline{v}'\, \underline{\theta}_1, \\ &= +1, \text{ if } u > \underline{v}'\, \underline{\theta}_1. \end{aligned}$$

We now perturb $f(\underline{x})$ by some distribution $g(\underline{x})$ and consider the corresponding perturbation on $\underline{\theta}_1$. Let the perturbation be

$$f^*(\underline{x}) = (1 - t)\, f(\underline{x}) + t\, g(\underline{x}), \quad 0 \leq t \leq 1$$

and let it produce the perturbed solution $\underline{\theta}_1^*$ according to

$$\int \underline{v}\, \psi(u - \underline{v}'\, \underline{\theta}_1^*)\, f^*(\underline{x})\, d\underline{x} = 0.$$

We reorganize this integral equation in order to obtain $\underline{\theta}_1^*$ as a function of $\underline{\theta}_1$ and $g(\underline{x})$. We immediately have the implicit equation

$$(1 - t) \int \underline{v} [\psi(u - \underline{v}'\, \underline{\theta}_1^*) - \psi(u - \underline{v}'\, \underline{\theta}_1)]\, f(\underline{x})\, d\underline{x}$$

$$+\, t \int \underline{v}\, \psi(u - \underline{v}'\, \underline{\theta}_1^*)\, g(\underline{x})\, d\underline{x} = 0.$$

The integrand of the first term, in fact, is non-zero in a very limited domain; and the integral is equal to

$$- 2 \int \underline{v}\ \psi(u - \underline{v}'\ \underline{\theta}_1)\ f(\underline{y})\ d\underline{y}$$

with

$\underline{y} = \underline{x}$ "between" the two hyperplanes
$\underline{u} = \underline{v}'\ \underline{\theta}_1$ and $u = \underline{v}'\ \underline{\theta}_1^*$

and is thus continuous with respect to $\underline{\theta}_1^*$. The second term of the implicit equation is relative to $g(\underline{x})$, an unspecified distribution so far. We will assume that $g(\underline{x})$ is also sufficiently regular and will later discuss that condition. Then we have obtained that both terms are varying smoothly with $\underline{\theta}_1^*$. But both are vectorial and their sum can only cancel for a discrete set of values $\underline{\theta}_1^*$, when t and $g(\underline{x})$ are arbitrarily given. Moreover $\underline{\theta}_1^*$ is continuous with respect to t and we know the trivial solution $\underline{\theta}_1^* = \underline{\theta}_1$ for $t = 0$. This is enough to conjecture the validity of the Taylor-like expansion

$$\underline{\theta}_1^* = \underline{\theta}_1 + t\ \underline{a}_1 + (t^2/2)\ \underline{a}_2 + \ldots$$

when t is small, that is when $\underline{\theta}_1^*$ remains in the vicinity of $\underline{\theta}_1$. Strict demonstration would be dependent upon the regularity conditions needed to apply the implicit function theorem.

What are coefficients $\underline{a}_1$, $\underline{a}_2$, ... comes out particularly easily in the one-dimension problem, $m = 1$, $v = 1$. Then the median θ of the distribution $f(x)$ satisfies

$$\int_{-\infty}^{+\infty} \psi(x - \theta)\ f(x)\ dx = 0,$$

whereas we are interested by its relation with θ^* in

$$\int_{-\infty}^{+\infty} \psi(x - \theta^*)\ [\,(1 - t)\ f(x) + t\ g(x)]\ dx = 0.$$

We develop as we did previously to obtain

$$(1 - t)\ 2 \int_{\theta^*}^{\theta} f(y)\ dy + t \int_{-\infty}^{+\infty} \psi(x - \theta^*)\ g(x)\ dx = 0$$

or, approximately,

$$(1 - t)\ 2\ (\theta - \theta^*)\ f(\theta) + t \int_{-\infty}^{+\infty} \psi(x - \theta)\ g(x)\ dx = 0$$

and

$$\theta^* = \theta + t \int_{-\infty}^{+\infty} \frac{\psi(x - \theta)}{2f(\theta)} g(x)\, dx + \frac{t^2}{2} \ldots$$

Whether $g(\underline{x})$ has to be differentiable everywhere is a partly open question. On the one hand, the limiting scheme, $\nu \to 1$, or an approximation of $g(x)$ by a differentiable function, is sufficient to allow the above presentation. On the other hand, a fairly rough function $g(x)$ may mean a very poor convergence of the Taylor-like expansion. We support these remarks.

The critical point of the derivation is the continuity of the second term in the implicit equation with respect to $\underline{\theta}_1^*$. When a discontinuous $g(\underline{x})$ is met with, the integral

$$\int \underline{v}\, \psi(u - \underline{v}'\, \underline{\theta}_1^*)\, g(\underline{x})\, d\underline{x},$$

seen as a function of $\underline{\theta}_1^*$, is discontinuous at each $\underline{\theta}_1^*$ such that

$$u = \underline{v}'\, \underline{\theta}_1^*$$

with

$$\underline{x} = (u, \underline{v}')' \text{ : discontinuity of } g(\underline{x}).$$

Note the discontinuities of the integral provide a partition of the parameter space by a set of hyperplanes - To allow the presentation, we have smoothed these discontinuities; but they may be intrinsically present.

Consider now variation of the parameter t, from $t = 0$ to $t = 1$. That may mean that $\underline{\theta}_1^*$ moves through several connex parts of the above partition and thus has a fairly discontinuous variation. This can only be expressed by a slowly converging expansion. When $\underline{\theta}_1^*$ remains in the neighbourhood of $\underline{\theta}_1$, that is in the same part of the partition, the expansion can converge very rapidly and therefore be truncated at the level of its first order term.

4.3.4. "Best" robust location estimator

We design in this section the function $\rho_1(\varepsilon)$, or $\psi(\varepsilon)$, which yields a *minimum asymptotic variance* given that

- the distribution $f(u)$ is known, except for its location, and continuous, for $u \in [u_-, u_+]$. Moreover $f(u_-) = f(u_+) = 0$.
- the influence curve is everywhere bounded (criterion of robustness).

- the solution θ is uniquely defined (criterion of convexity).

Although only location estimation is concerned in the sequel, the same type of argument could be produced in multiple regression estimations. However, this would be to the expense of a fairly involved investigation of the relationships existing between the distributions of u, $\underline{v}$ and the residuals ε; we have not thought this effort profitable seeing the limited results we already have.

Thus, we intend to minimize the asymptotic variance

$$V = \sigma_1^2 = \frac{\int [\psi(u - \theta)]^2 f(u)\, du}{[\int \phi(u - \theta)\, f(u)\, du]^2}$$

by an appropriate choice of the function $\psi(\varepsilon)$. We first retain the attention on the fact that, for the best, the function $\psi(\varepsilon)$ can only be defined up to some multiplicative factor. The latter will be selected in order to have a unit denominator.

In way of introduction to the derivation, we first omit the constraints of robustness and convexity. We therefore fit more or less in the frame of Huber's paper (1964) when he investigates his minimax questions. However the approach will be quite different in many respects.

In order to ease the analytical manipulations, we will use throughout this section a vectorial-matricial notation resulting from a discretisation of the sample space. Functions become vectors, whereas operators are matrices.

Let the space coordinate be discrete. Any function y(u) will then be defined by the set of values $\{\ldots, y(u_{i-1}), y(u_i), y(u_{i+1}), \ldots\}$ taken at the regularly spaced coordinates $\{\ldots, u_{i-1}, u_i, u_{i+1}, \ldots\}$ with

$$u_j = u_0 + j\eta, \quad \eta \text{ infinitesimal.}$$

- Remark the representation basis will only be an intermediate step and is not essential, other representations can be preferred - To the set of values, we associate a vector $\underline{y}$ of possibly infinite dimension.

We will assume that functions are sufficiently differentiable, e.g.,

$$\text{to } (\partial/\partial u)\, y(u), \text{ we associate } D\underline{y}$$

where the matrix D has elements

$$d_{ij} = \begin{cases} 1/(2\eta), & \text{if } i = j-1 \\ -1/(2\eta), & \text{if } i = j+1 \\ 0, & \text{otherwise.} \end{cases}$$

When $\underline{y}$ does not vanish at the frontier of the open set of differentiation, difficulties occur which will be taken care of by similarity with the ordinary infinitesimal calculus.

We also need an integration operator. It will be of matricial type in order to support an operation with respect to a weight f(u) and two arbitrary functions y(u) and z(u), e.g.,

to $\int y(u)\, z(u)\, f(u)\, du$, we associate $\underline{y}'\, F\, \underline{z} = \underline{z}'\, F\, \underline{y}$

where the matrix F is diagonal and has elements

$$F = \eta \operatorname{diag}(\ldots, f(u_i), \ldots).$$

When only one function is concerned, the unit constant is substituted to the other, e.g.,

to $z(u) = 1$, we associage $\underline{1}$,
to $\int y(u)\, f(u)\, du$, we associate $\underline{y}'\, F\, \underline{1}$

With this formalism, we transform the function analysis problem in a standard minimization with equality and inequality constraints. At the moment, we only take into account the equalities.

Let $\underline{y}$ be the vector associated to the $\psi(u - \theta)$, then it minimizes

$$V = \underline{y}'\, F\, \underline{y}$$

under constraints

$$c_1 = 2\, \underline{1}'\, F\, \underline{y} = 0$$

and

$$c_2 = 2(\underline{1}'\, F\, D\, \underline{y} - 1) = 0.$$

Constraint c_1 indicates that θ is an M-estimator, whereas constraint c_2 stands for the denominator of the asymptotic variance.

We solve this minimization by the method of the Lagrange multipliers; let them be λ_1, λ_2. Then, $\underline{y}$ is also minimum of

$$V + \lambda_1 c_1 + \lambda_2 c_2$$

or, after differentiation with respect to $\underline{y}$,

$$F \underline{y} + \lambda_1 F \underline{1} + \lambda_2 D' F \underline{1} = 0$$

and

$$\underline{y} = - \lambda_1 \underline{1} + \lambda_2 F^{-1} DF \underline{1}.$$

The last transformation makes use of the antisymmetry of matrix D, and can be performed only if the distribution vanishes on the frontier of its domain of definition; i.e.

$$D'F = - DF \Leftrightarrow f(u_-) = f(u_+) = 0.$$

Introduction of $\underline{y}$ in c_1 and c_2 provides the Lagrange multipliers

$$\lambda_1 = \lambda_2 \underline{1}' D F \underline{1}, \text{ given } \underline{1}' F \underline{1} = 1,$$
$$\lambda_2 = 1/\underline{1}' F D F^{-1} D F \underline{1}.$$

The former cancels accordingly with

$$\begin{aligned} \lambda_1/\lambda_2 &= \underline{1}' D F \underline{1} \\ &= \int (\partial/\partial u) f(u) du \\ &= f(u_+) - f(u_-) = 0. \end{aligned}$$

We collect the results and obtain

$$\psi(u - \theta) = \lambda_2 f(u)^{-1} (\partial/\partial u) f(u)$$

or

$$\psi(u - \theta) = \lambda_2 (\partial/\partial u) \ln f(u).$$

Given c_1, i.e.

$$\int \psi(u - \theta) f(u) du = 0,$$

we see that any M-estimator which is equal to the maximum likelihood estimator, except for a translation constant, is admissible in the sense of Stein (1955), when the distribution vanishes on the sample space boundary. The asymptotic variance of this M-estimator is given by

$$\begin{aligned} V &= \underline{y}' F \underline{y} \\ &= - \lambda_2 \end{aligned}$$

where

$$\lambda_2 = 1 / \int [(\partial^2/\partial u^2) \ln f(u)] f(u) du.$$

Although the characteristics of maximum likelihood estimators are classical, we illustrate the above findings in order to display the relations between the various elements. Let us estimate the location of the sample $u_1,\ldots,u_n$ drawn from the gamma distribution of density

$$f(u) = [1/\Gamma(\nu + 1)]\ (u - a)^{\nu} \exp[-(u - a)],\ \text{if } u > a$$
$$= 0 \qquad\qquad\qquad\qquad\qquad\qquad\qquad\qquad\qquad ,\ \text{if } u < a$$

for strictly positif parameter ν.

Note that the restriction on parameter ν is peremptory. Effectively, we have derived the minimum variance M-estimator with respect to a differentiable distribution vanishing on the sample space boundary. In the present case, it is convenient to set

$$u_- = a \text{ and } u_+ = \infty.$$

This yields

$$\psi(u - \theta) = \lambda_2\ \{[\nu/(u - a)] - 1\}$$

and

$$V = -\lambda_2 = \nu - 1,\ \text{for } \nu > 1.$$

Therefrom, and accordingly with

$$\int \psi(u - \theta)\ f(u)\ du = 0,$$

we can define θ through

$$\sum w_i\ \{[\nu/(u_i - \theta)] - 1\} = 0$$

or by the explicit result derived with respect to θ_0, an approximation of θ.

$$\theta = \theta_0 + \sum w_i\ [(1/\nu) - 1/(u_i - \theta_0)]\ /\ [\sum w_i/(u_i - \theta_0)^2].$$

This estimator is minimum in asymptotic variance but obviously is not robust. - By the way, note that V tends to cancel for $\nu = 1$; this is indicative of difficulties in the analytical conditions which lead the expansion of section 2.4 to converge too slowly. Then the variance of θ is not anymore inversely proportional to the sample size n; the high order terms dominate the first in the expansion

$$n\ \mathrm{var}(\theta) = V + O(n^{-1}).$$

We now proceed in the derivation by addition of the inequality constraints which impose robustness to the M-estimator. We search for a minimum variance θ and thus have to minimize

$$V = \underline{y}' F \underline{y}$$

under equality constraints

$$c_1 = 2 \underline{1}' F \underline{y} = 0$$

and

$$c_2 = 2(\underline{1}' F D \underline{y} - 1) = 0.$$

We further limit $\rho(.)$ to be convex, that is we restrict by

$$c_{3k} = - 2\underline{e}'_k D \underline{y} < 0,$$

for any basis vector $\underline{e}_k = (0,...,0,1,0,...,0)'$. We also set a superior bound to the influence curve. This bound will be set relative to the mean quadratic value. Thus a second group of inequality constraints may be stated in the form

$$c_{4l} = (e'_l \underline{y})^2 - \beta \underline{y}' F \underline{y} < 0.$$

Inspection reveals that the set of equations has a non trivial solution if, and only if,

$$\beta > 1.$$

The situation $\beta = 1$, seen as the limit of greater β-values, gives the median of $f(u)$ for solution. It is the most robust θ.

We investigate this inequality constrained minimization by the method of Kuhn and Tucker, according to Beveridge and Schechter (1970, section 4.3.3). Similarly to the above derivation, the solution $\underline{y}$ must be minimun of

$$V + \lambda_1 c_1 + \lambda_2 c_2 + \sum \lambda_{3k} c_{3k} + \sum \lambda_{4l} c_{4l},$$

but the inequalities lead the last two sums to satisfy

$$\sum \lambda_{3k} c_{3k} + \sum \lambda_{4l} c_{4l} = 0,$$

with the following constraints on the Lagrange multipliers

$$\lambda_{3k} > 0 \quad \text{and} \quad \lambda_{4l} > 0$$

for all k and l. There is a further requirement concerning the region delimited by the inequalities; the minimum must be accessible and thus not exterior to this region. This question of accessibility will be considered as resolved seeing that $\beta > 1$ guarantees the existence of an accessible minimum.

Consider now the condition

$$\sum \lambda_{3k} c_{3k} + \sum \lambda_{4l} c_{4l} = 0.$$

Each term can only contribute in a non positive way, therefrom we conclude that, at the accessible minimum, we necessarily have

$$\begin{array}{lll} \text{either } \lambda_{3k} = 0 & & \text{and } c_{3k} < 0 \\ \text{or } \quad \lambda_{3k} > 0 & & \text{and } c_{3k} = 0, \end{array}$$

as well as,

$$\begin{array}{lll} \text{either } \lambda_{4l} = 0 & & \text{and } c_{4l} < 0 \\ \text{or } \quad \lambda_{4l} > 0 & & \text{and } c_{4l} = 0. \end{array}$$

The meaning of these alternatives is obviously that, for the first terms, the constraints are not binding and that, for the second terms, the minimum lies on the boundary of the accessibility region. In the sequel, we will carry the attention on the second terms; this is in the line of the thorough discussion of the Kuhn-Tucker approach given by Vajda (1961, section 12.4).

We differentiate the combined expression, with respect to $\underline{y}$, in order to obtain the minimum. This yields to

$$(1 - \beta \sum \lambda_{4l}) F \underline{y} + \lambda_1 F \underline{1} + \lambda_2 D' F \underline{1}$$
$$- \sum \lambda_{3k} D' \underline{e}_k + \sum \lambda_{4l} (\underline{e}'_l \underline{y}) \underline{e}_l = 0$$

and

$$\underline{y} = (1 - \beta \sum \lambda_{4l})^{-1} [- \lambda_1 \underline{1} + \lambda_2 F^{-1} D F \underline{1}$$
$$- \sum \lambda_{3k} F^{-1} D \underline{e}_k - \sum \lambda_{4l} (\underline{e}' \underline{y}) F^{-1} \underline{e}_l].$$

This is under conditions

$$D' F = - DF \text{ or } f(u_-) = f(u_+) = 0$$

and

$$\lambda_{3k} D' \underline{e}_k = - \lambda_{3k} D \underline{e}_k .$$

The latter implies that the convexity criterion cannot be constraining on the frontier of the sample space (λ_{3k} must cancel on the frontier).

The expression obtained for $\underline{y}$ in terms of the Lagrange multipliers will now be analysed. First, note that the multiplicative scalar factor $(.)^{-1}$ can be omitted, if we accept new definitions of the Lagrange multipliers. Second, note that matrices F^{-1} and $(F^{-1}D)$ are, or nearly are, diagonal. And third, assume that $f(u)$ has a continuous first derivative in order to obtain a continuous $\psi(u - \theta)$. Then, three different types of behaviours of $\psi(u - \theta)$ can be distinguished in connex subsets of the sample space. They are

- Type 1 : the two groups of inequalities are not binding. Then we obtain in corresponding subsets.

$$\lambda_{3k} = \lambda_{4l} = 0$$
$$\underline{y} = \lambda_2 F^{-1} D F \underline{1} - \lambda_1 \underline{1}$$

or

$$\psi(u - \theta) = \lambda_2 [(\partial/\partial u) \ln f(u)] - \lambda_1$$

with

$$[\psi(u - \theta)]^2 \leq \beta V$$

and

$$\phi(u - \theta) \geq 0.$$

- Type 2 : in subsets where the convexity criterion is binding, we have

$$c_{3k} = 0, \; c_{4l} < 0$$

and

$$\psi(u - \theta) = \text{constant}$$

with

$$[\psi(u - \theta)]^2 < \beta V.$$

- Type 3 : in the remaining subsets of the sample space, the robustness criterion is binding. Thus

$$c_{41} = 0, \quad c_{3k} < 0$$

and

$$\psi(u - \theta) = \pm \sqrt{\beta V} .$$

Observe that $\psi(u - \theta)$, has, at most in two subsets, a type 3 behaviour.

The above derivation fully supports the conjecture in Huber (1972, section 12.3) concerning robust maximum likelihood estimators. But, moreover, it demonstrates how to estimate minimum variance robust M-estimators in the present location case.

To illustrate the above discourse, we consider robust estimation for the previously met example. We estimate the location of the sample $u_1,\dots,u_n$ drawn from the gamma distribution and we require that all observations contribute in a similar way to the estimate. In more precise terms, we may require that the maximum incidence be twice the "quadratic mean" or $\beta = \beta_0$ with $\beta_0 = 2^2 = 4$.

We now determine the equation of $\psi(u - \theta)$. First, let us find its structure. There is certainly a type 1 subset, given $\beta > 1$. There, $\psi(u - \theta)$ has the expression, written with unknown coefficients b_1 and b_2,

$$\psi(u - \theta) = b_1 [1/(u - a) - b_2].$$

Seeing that the convexity criterion is never binding, we conclude to the existence of a unique type 1 subset and to the absence of type 2. There can be one or two subsets of type 3. Noting b_1b_3 the corresponding bound, we summarize the structure of $\psi(u - \theta)$ as follows

$$\begin{aligned} \psi(u - \theta) &= - b_1b_3, && \text{if } u - a < b_4 \\ &= b_1[1/(u - a) - b_2], && \text{if } b_4 < u - a < b_5 \\ &= + b_1b_3, && \text{if } u > b_5 \end{aligned}$$

with

$$\begin{aligned} & b_1, b_2, b_3 > 0 \\ & b_4 = 1/(b_2 + b_3) \\ & b_5 = 1/(b_2 - b_3), && \text{if } b_2 > b_3 \\ & \quad = \infty, && \text{if } b_2 < b_3 . \end{aligned}$$

- It may be noteworthy to observe that this is very nearly the result obtained by Huber (1964, section 6) for the contaminated normal distribution although the present context differs.

The way we have parametrized the function $\psi(u - \theta)$ permits an easy determination of its coefficients. In order to produce a minimum variance robust M-estimator, the coefficients must be such that they verify the constraint c_1, i.e.

$$\int_a^\infty \psi(u - \theta)\, f(u)\, du = 0,$$

such that they satisfy the constraint c_2, i.e.

$$\int_a^\infty [(\partial/\partial u)\, \psi(u - \theta)]\, f(u)\, du = 1,$$

as well as such that they satisfy c_{41} in the type 3 subsets, i.e.

$$\max[\psi(u - \theta)]^2 = (b_1 b_3)^2 = \beta V$$

with

$$V = \int_a^\infty [\psi(u - \theta)]^2\, f(u)\, du.$$

Due to the absence of type 2 subset, we have omitted the inequality constraints c_{3k}.

This set of implicit equations can easily be solved by numerical means. We propose in Table 1 the results obtained for a few particular values of ν and β. Our concern is for $\beta = 4$ and we see that the robust θ exhibits good efficiency. With this β-value, its asymptotic variance is approximately $V = \nu$; it is more efficient than the non-robust arithmetic mean,

$$\text{relative efficiency} = \nu/(\nu + 1),$$

but less efficient than the non-robust minimum variance M-estimator,

$$\text{relative efficiency} = \nu/(\nu - 1).$$

To further illustrate, we write down the solution for the parameter set $\beta = 4$, $\nu = 3$. Accordingly with

$$\int \psi(u - \theta)\, f(u)\, du = 0,$$

we define θ through the implicit equation

$$\sum w_i \, \psi(u_i - \theta) = 0$$

or

$$.2849 \sum w_j - \sum w_k [1/(u_k - \theta) - .3076] = 0$$

where

$$u_i = u_j, \text{ if } u_i < 1.688 + \theta$$

$$= u_k, \text{ if } u_i > 1.688 + \theta .$$

Some 9 percents of the weights lie in the type 3 subset; there the observations are taken into account through their weights and irrespectively of their exact values.

		b_1	b_2	b_3	V	$\int_{b_4}^{b_5} f(u)\,du$
$\nu=2$,	$\beta=4$	6.002	.4401	.4698	1.988	.9006
	$\beta=9$	4.114	.4694	.9171	1.582	.9632
	$\beta=25$	3.052	.4879	1.878	1.314	.9908
	$\beta=100$	2.458	.4968	4.353	1.145	.9988
$\nu=3$,	$\beta=4$	12.07	.3076	.2849	2.956	.9086
	$\beta=9$	8.921	.3219	.5309	2.492	.9686
	$\beta=25$	7.212	.3297	1.031	2.212	.9932
	$\beta=100$	6.369	.3327	2.256	2.065	.9993
$\nu=4$,	$\beta=4$	20.27	.2360	.1959	3.942	.9144
	$\beta=9$	15.68	.2444	.3546	3.435	.9723
	$\beta=25$	13.30	.2485	.6675	3.154	.9947
	$\beta=100$	12.28	.2498	1.418	3.032	.9996
$\nu=10$,	$\beta=4$	114.1	.0980	.0553	9.958	.9303
	$\beta=9$	97.92	.0994	.0934	9.295	.9826
	$\beta=25$	91.42	.0999	.1645	9.046	.9981
	$\beta=100$	90.06	.1000	.3331	9.002	1.000
$\nu > 1$,	β large	$\nu(\nu-1)$	$1/\nu$	*	$\nu-1$	

* $b_3 = \beta^{1/2} [\nu^2(\nu-1)]^{-1/2}$

Table 1

The estimator θ is consistent with respect to parameter a. It may be applied to the assessment of the location of any sample distributed approximately according to f(u). However, what θ means for other distributions than f(u) may present conceptual difficulties.

4.4. MM-estimators in regression estimation

We have observed at section 4.3.1 that it was not possible to obtain robust M-estimators of regression without involving a dependence on some "scale" parameter - This is true even in the simple point location - Therefore, we are compelled to include a scale estimation whenever we solve robustly a regression problem.

We will be mainly concerned by the estimation of the scale of the residuals but we could as well be in need of some multidimensional scatter estimation. This is the object of the investigation of Maronna (1976) who defines simultaneously location and scatter by a set of MM-estimators.

But what is the "scale" of the residuals ? A partially disappointing answer has been proposed by Huber (1964). We excerpt :

> "The theory of estimating a scale parameter is less satisfactory than that of estimating a location parameter. Perhaps the main source of trouble is that there is no natural "canonical" parameter to be estimated. In the case of a location parameter, it was convenient to restrict attention to symmetric distributions; then there is a natural location parameter, namely the location of the center of symmetry, and we could separate difficulties by optimizing the estimator for symmetric distributions (where we know what we are estimating) and then investigate the properties of this optimal estimator for nonstandard conditions, e.g., for nonsymmetric distributions. In the case of scale parameters, we meet, typically, highly asymmetric distributions, and the above device to ensure unicity of the parameter to be estimated fails. Moreover, it becomes questionable, whether one should minimize bias or variance of the estimator."

The same author has recently investigated various approaches to scale and scatter definitions (1977a). For our part, at the moment we will only assume that, given a set of residuals $\varepsilon_1,\dots,\varepsilon_n$ corresponding to some estimator $\underline{\theta}_1$ through

$$\varepsilon_i = u_i - \underline{v}'_i \, \underline{\theta}_1,$$

we have a minimization rule which provides the scale s according to

$$s = \theta_2$$

and

$$M_2 = \min \text{ for } \theta_2$$

with

$$M_2 = \int \rho_2 \, (\underline{x}, \underline{\theta}_1, \theta_2) \, f(\underline{x}) \, d\underline{x}.$$

In order to provide a measure of the scale of the residuals, the function $\rho_2(.)$ must be selected such that

$$\theta_2 = \text{scale of } (\varepsilon_1, \ldots, \varepsilon_n)$$

is consistent with

$$|\lambda| \theta_2 = \text{scale of } (\lambda \, \varepsilon_1, \ldots, \lambda \, \varepsilon_n), \; \lambda \in R.$$

We now devote our attention to the MM-estimator $\underline{\theta}_1$. It will be such that

$$M_1 = \min \text{ for } \theta_1$$

with

$$M_1 = \int \rho_1 \, (\underline{x}, \underline{\theta}_1, \theta_2) \, f(\underline{x}) \, d\underline{x}$$

where ρ_1 is constrained to yield compatibility between

$$\underline{\theta}_1 = \text{regression estimate on } (\underline{x}_1, \ldots, \underline{x}_n)$$

and

$$\underline{\theta}_1 = \text{regression estimate on } (\lambda \, \underline{x}_1, \ldots, \lambda \, \underline{x}_n), \; \lambda \in R_+.$$

The two conditions on the ρ_1- and ρ_2-structures constrain the former to have the very natural form

$$\begin{aligned} \rho_1(\underline{x}, \underline{\theta}_1, \theta_2) &= \rho_1(\underline{x}/\theta_2, \underline{\theta}_1) \\ &= \rho_1 \, [(u - \underline{v}' \, \underline{\theta}_1)/\theta_2] \\ &= \rho_1 \, (\varepsilon/s). \end{aligned}$$

In the two next subsections, we present the main computational procedures in use to solve the set of implicit equations

$$\int \psi_1(\underline{x}, \underline{\theta}_1, \theta_2)\ f(\underline{x})\ d\underline{x} = 0,$$
$$\int \psi_2(\underline{x}, \underline{\theta}_1, \theta_2)\ f(\underline{x})\ d\underline{x} = 0.$$

The last subsection indicates the main proposals designed in order to provide robust estimators.

4.4.1. Relaxation methods

Although many variations are possible in these methods, we will only present the main algorithm where no damping factor appears. This will be sufficient to indicate where lie the difficulties.

The algorithm is iterative and consists in a treatment which is repeated one or several times until the convergence is deemed sufficient. Each repetition, or step, will be identified by an index k and we will assume that we have at disposal $\underline{\theta}_1^k$ and θ_2^k, approximations of $\underline{\theta}_1$ and θ_2, when the step begins. Initially, we can enter the solution corresponding to the least squares method.

In step k, we start with the approximate solution $(\underline{\theta}_1^k, \theta_2^k)$ and we improve in $(\underline{\theta}_1^{k+1}, \theta_2^{k+1})$ by the scheme

- Compute θ_2^{k+1}, the solution of

$$\int \psi_2(\underline{x}, \underline{\theta}_1^k, \theta_2^{k+1})\ f(\underline{x})\ d\underline{x} = 0 .$$

- Compute $\underline{\theta}_1^{k+1}$, the solution of

$$\int \psi_1(\underline{x}, \underline{\theta}_1^{k+1}, \theta_2^{k+1})\ f(\underline{x})\ d\underline{x} = 0 .$$

In principle these two equations could be difficult to solve, but in practice no great difficulty is encountered at this level. The first has usually an explicit solution

$$s = \theta_2^{k+1} = S(\varepsilon_1^k, \ldots, \varepsilon_n^k)$$

where S(.) is a known function of the residuals. The second equation can be reorganized in order to be solved by any least squares regression algorithm. We detail further.

First observe that, θ_2^{k+1} being already estimated, we are estimating an M-estimator of regression (g = 1). Therefore, we will join the notation of section 4.3. Let us define the scalar function $\psi(\varepsilon)$ as we did previously, i.e. according to

$$\underline{\psi}_1(\underline{x}, \underline{\theta}_1, \theta_2) = \underline{\psi}_1 [(u - \underline{v}'\ \underline{\theta}_1)/\theta_2]$$
$$= - \psi(u - \underline{v}'\ \underline{\theta}_1)\ \underline{v}$$
$$= - \psi(\varepsilon)\ \underline{v},$$

then the equation can be written

$$\sum w_i\ \psi(\varepsilon_i^k)\ \underline{v}_i = 0$$

for

$$f(\underline{x}) = (1/\sum w_i) \sum w_i\ \delta(\underline{x} - \underline{x}_i).$$

And, to conclude, the equation is transformed in

$$\sum w_i [\psi(\varepsilon_i^k)/\varepsilon_i^k]\ (u_i - \underline{v}'_i\ \underline{\theta}_1)\ \underline{v}_i = 0,$$

or

$$\{\sum w_i [\psi(\varepsilon_i^k)/\varepsilon_i^k]\ \underline{v}_i\ \underline{v}'_i\}\ \underline{\theta}_1 = \sum w_i [\psi(\varepsilon_i^k)/\varepsilon_i^k]\ u_i\ \underline{v}_i;$$

this produces an improved estimate $\underline{\theta}_1$ by inversion of the last linear system. Whether effectively an improvement results depends upon the nature of the function $\psi(\varepsilon)$. It may easily be seen that $\underline{\theta}_1$ converges to $\underline{\theta}_1^{k+1}$, by repetition of the last computation, whenever $\psi(\varepsilon)$ is admissible in the sense of section 4.3.1. Note that the above approach is quite general and that, in specific situations, well selected methods can be very much cheaper in computation time - see Huber (1973, section 8) as well as Huber and Dutter (1974).

An important drawback of this relaxation method is that it is not clear whether it converges. It may be observed that sometimes the starting set $(\underline{\theta}_1^0, \theta_2^0)$ has the utmost importance on the final solution. Investigation of a few pathological situations has revealed to this author that this can be associated to the existence of several minima (and possibly saddle-points) in the parameter space of $(\underline{\theta}_1, \theta_2)$ - This problem will be assessed at section 4.5 - For the time being we assume that, after possibly a few erratic steps, the relaxation method does converge. Thus, in the vicinity of the final solution, the convergence is linear and corresponds approximately to

$$\underline{\theta}_1^{k+1} - \underline{\theta}_1^k = A_{11}^{-1}\ A_{12}\ A_{22}^{-1}\ A_{21}\ (\underline{\theta}_1^k - \underline{\theta}_1^{k+1}),$$

where

$$A_{jk} = \int \phi_{jk}\ (\underline{x}, \underline{\theta}_1, \theta_2)\ f(\underline{x})\ d\underline{x}.$$

We omit the argument leading to this result seeing its similarity with the derivation presented at section 4.2.

In spite of the obvious deficiencies of the above approach and frequently without knowledge of the possible hazards of convergence, most experimenters have adopted a relaxation method. This is understandable when taking into consideration that most computer centers have at disposal an efficient software to solve generalized least squares regression problems. The algorithms are generally denoted as "reweighted" least squares. Recent experience has been reported by Gross (1977) as well as Ypelaar and Velleman (1977), a software package is presented by Coleman et al. (1977).

4.4.2. Simultaneous solutions

Various methods have been proposed to estimate simultaneously the two sets of values $\underline{\theta}_1$ and θ_2. We will describe a fairly general method at section 4.5 based on a modification of the problem at hand, but, at the moment, we only consider the direct solution by iterations which simultaneously produce improved estimates of $\underline{\theta}_1$ and θ_2. In fact, we perform Newton-Raphson iterations in the space of $(\underline{\theta}_1, \theta_2)$.

The argument is quite general and can be presented for any number g of equations defining an MM-estimator. It presents many similarities with the derivation of section 4.2 leading to an expression of the influence function $\Omega_j(x)$.

Assume we have at disposal a set of approximations $(\theta_1^*,\ldots,\theta_g^*)$, and that we want the solution $(\theta_1,\ldots,\theta_g)$ of the set of g equations

$$\int \psi_j \, fdx = \int \psi_j(x, \theta_1,\ldots, \theta_g) \, f(x) \, dx = 0.$$

Each equation is approximately satisfied by the set $(\theta_1^*,\ldots, \theta_g^*)$ and we devote our attention only to the first order terms in the following expansions. Under the usual conditions of differentiability, we have

$$\begin{aligned} \int \psi_j^* \, fdx &= \int [\psi_j + \sum \phi_{jk} (\theta_k^* - \theta_k)] \, fdx \\ &= \int \psi_j \, fdx + \sum [\int \phi_{jk} \, fdx] \, (\theta_k^* - \theta_k) \\ &= \sum A_{jk}(\theta_k^* - \theta_k). \end{aligned}$$

We see the difference $(\theta_j^* - \theta_j)$ is the solution of a set of g linear equations. When the coefficients A_{jk} $(k \neq j)$ are dominated by A_{jj}, that is when the estimators are relatively independent from one another, a solution can be derived. Under

$$\| \sum_k A_{ik} A_{kk}^{-1} A_{kj} A_{jj}^{-1} \| << 1, \; k \neq i, \; k \neq j,$$

we obtain

$$\theta_j = \theta_j^* - \{\sum A_{jj}^{-1} A_{jk} A_{kk}^{-1} \int \psi_k^* \, fdx - 2 A_{jj}^{-1} \int \psi_j^* \, fdx\}$$

$$= \theta_j^* - \int \Omega_j^* \, fdx$$

$$= \theta_j^* - \int \Omega_j \, (x, \theta_1^*, \ldots, \theta_g^*) \, f(x) \, dx .$$

Inspection of the mathematical treatment reveals that the last expression is correct even if the estimators are strongly dependent upon one another.

Therefrom we conclude that robust MM-estimators are obtained through finite increments $\theta_j - \theta_j^*$. This is important in view of the next section where Ω_j becomes a continuous function of some parameter independent from $(\theta_1, \ldots, \theta_g)$. Further if we observe large increments, this must lead us to suspect lack of robustness of the concerned MM-estimator.

4.4.3. Some proposals

Basically very few proposals have been advanced and their level of robustness is frequently difficult to appreciate. The question is what functions ψ_1 and ψ_2 should we select in

$$\int \psi_1(x, \underline{\theta}_1, \theta_2) \, f(\underline{x}) \, d\underline{x} = 0$$

and

$$\int \psi_2(x, \underline{\theta}_1, \theta_2) \, f(\underline{x}) \, d\underline{x} = 0$$

in order to obtain robust estimate of the regression $\underline{\theta}_1$ in the model

$$u = \underline{v}' \, \underline{\theta}_1 + \varepsilon$$

and robust scale estimate of the residuals

$$s = \theta_2 = \text{scale of } \varepsilon .$$

- Let us recall that both are simultaneously robust or non-robust - Possibly the most exhaustive set of proposals has been conceived by the Princeton Robustness Group (see Andrews et al., 1972) and it has been thoroughly described by Gross and Tukey (1973).

Although the Princeton Study has been largely concerned by location problems, many of their estimators can be adapted to the regression field. In this section, we disregard the computational procedures although they should be well adapted to the problem at hand. Thus, we will not retain the attention on the various one-step estimators which are obtained through a single iteration of the relaxation method seen at section 4.4.1. Further details can be drawn from Bickel (1975) as well as in the already referred work.

There is no obvious reason to associate a given selection of $\psi_1(.)$ to a specific selection of $\psi_2(.)$, although this is frequently performed. At the moment, we dissociate these selections.

Assume that for known scale factor s, we estimate $\underline{\theta}_1$ through

$$M_1 = \sum w_i \, \rho_1(\varepsilon_i) = \min \text{ for } \underline{\theta}_1$$

with

$$\varepsilon_i = u_i - \underline{v}'_i \, \underline{\theta}_1,$$

then the two main classes of proposals are the following : first, a family due to Huber

$$\begin{aligned} \rho_1(\varepsilon) &= \varepsilon^2, & \text{if } |\varepsilon| \leq ks \\ &= ks\ (2\ |\varepsilon| - ks), & \text{if } |\varepsilon| \geq ks, \end{aligned}$$

which is a parabola prolonged by two tangents and, second, families of functions becoming gradually constant for large residuals, among them a proposal due to Andrews

$$\begin{aligned} \rho_1(\varepsilon) &= 1 - \cos\,[\varepsilon/(cs)]\,, & \text{if } |\varepsilon| \leq \pi cs \\ &= 2 & \text{, if } |\varepsilon| \geq \pi cs. \end{aligned}$$

Both classes of proposals have bounded first and second derivatives whatever is ε and, for small residuals, are equivalent to the least squares method.

Investigating robust estimation in the context of contaminated normal distributions, Huber demonstrated the optimality of his proposal as being the minimax solution (1964) for the location problem with known scale parameter s. His proposal can be seen as a "robustified" least squares. It produces good results but possibly sub-optimal when the reference distribution is not normal.

In order to avoid any incidence of the outliers, Andrews (1974, 1975) has suggested to use a function $\rho(\varepsilon)$ constant for large residuals. This had also been considered by Hampel (proposals 12A to 25A in Andrews et al. 1972) and by Gross (1977), but, using a continuous $\psi(\varepsilon)$, these approaches yield to negative $\phi(\varepsilon)$ for some ε-domain. Therefore, these functions are not admissible and may produce unforseen results possibly in an unnoticed way. This will clearly appear in the illustration at the end of this section where, on the one hand, $\underline{\theta}_1$ exhibits discontinuities with respect to parameter c and, on the other hand, several solutions are possible for some given c.

We now turn our attention to the scale factor estimate. A seemingly robust definition is the median of the residuals in absolute values, i.e.

$$s = \theta_2 = \text{median}\ (|\varepsilon_1|, \ldots, |\varepsilon_n|)$$

or

$$\theta_2 = \lim \theta_{2\nu}, \text{ for } \nu \to 1,\ \nu > 1$$

in

$$M_2 = \sum w_i\ \rho_{2\nu}(\varepsilon_i) = \min \text{ for } \theta_{2\nu}$$

with

$$\rho_{2\nu}(\varepsilon) = |(|\varepsilon| - \theta_{2\nu})|^{\nu}.$$

This definition will be applied in the illustration. Several other definitions have been proposed frequently based on some order statistics (e.g., interquartile range) but they do not necessarily fit in the frame of section 2.4, furthermore they sometimes exhibit poor robustness. Some more insight on how to select a scale estimator can be gained by inspection of the tables in the appendix on contaminated normal distributions.

Once the two definitions of $\rho_1(\varepsilon)$ and $\rho_2(\varepsilon)$ have been decided upon, it remains to perform the computations leading to the eventual estimates; as seen, this may involve rather expensive calculations. In order to avoid this trouble, Huber and Dutter (1974) as well as Huber (1977b) propose to minimize another expression than (M_1, M_2) such that they obtain simultaneously $\underline{\theta}_1$ and θ_2. Their expression, resolved by Dutter (1977),

$$\sum w_i\ \rho[(u_i - \underline{v}'_i\ \underline{\theta}_1)/s]\ \ s + As = \min \text{ for } \underline{\theta}_1,\ s > 0$$

does not seem very appealing to us in spite of several favorable properties. We feel afraid by the level of arbitrariness.

Quite a different approach has already been mentioned, it consists in directly minimizing the scale estimate instead of M_1. This is, among alteri, the attitude of Jaeckel (1972b) and has been recommended by Hampel (1975) as providing the optimum breakdown point, but the argument appears dubious to us. When the definition of the scale factor is

$$s = \text{median } (|\varepsilon_1|, \ldots, |\varepsilon_n|),$$

the above procedure produces a $\underline{\theta}_1$-estimate corresponding to the selection of a function

$$\rho_1(\varepsilon) = \min{}_{[}(|\varepsilon|/s)^{\nu}, 1 + 1/\nu]$$

for very large power ν. It may, therefore, be seen as the limit of M-estimators. The above function is admissible, but not convex everywhere. The solution is not unique. Further, after Cover and van Campenhout (1977) it does not seem possible to design an algorithm able to escape the combinatorial complexity.

4.4.4. Illustration

We briefly report the observations made in comparing several methods of regression on a classical example.

The regression problem we are investigating has already been considered by many authors. It is relative to the operation of a plant for the oxydation of amonia into nitric acid and can be found in Brownlee (1965, section 13.12), Draper and Smith (1966, chapter 6), Daniel and Wood (1971, chapter 5), Andrews (1974) and Denby and Mallows (1977).

The data set has the size n = 21 and is of dimension p = 4. It consists in the regression of u_i, a stack loss, against v_{2i}, an air flow, against v_{3i}, a cooling water inlet temperature, as well as against v_{4i}, an acid concentration; the term $v_{1i} = 1$ introduces a constant in the regression

$$u_i = \underline{v}'_i \, \underline{\theta}_1 + \varepsilon_i.$$

In this example, various techniques have revealed that four observations (i = 1,3,4,21) are clearly outlying with respect to the distribution of the seventeen others which form a neat cluster.

Although this does not appear in the observation coordinates, this can be ascertained. A plot of the 21 observations is given in Figure 1 by a method of non-linear multidimensional scaling described in Rey (1976). This method produces a map of the 4-dimension space in a 2-dimension plane while preserving the distance relationships between observations; distant points remain distant and close points remain close. In order to be independent of the coordinate dimensionalities, the distance d_{ij} between two observations $\underline{x}_i$ and $\underline{x}_j$ has been defined according to

$$d^2_{ij} = \sum_k [(u_i - \underline{v}'_i \, \underline{\theta}_{1k}) - (u_j - \underline{v}'_j \, \underline{\theta}_{1k})]^2,$$

where $\underline{\theta}_{1k}$ is member of a set $\{\underline{\theta}_{1,1}, \underline{\theta}_{1,2}, \ldots\}$ of robust regression solutions. The exact composition of this set of solutions seems to have scarcely any incidence on the metric as long as they are independent.

Four different selections of the function

$$M_1 = \sum w_i \, \rho_1(\varepsilon_i)$$

to be minimized will now be considered and for each selection some results will be reported in Table 2, namely $\underline{\theta}_1$ and the corresponding scale factor s,

$$s = \text{median } (|\varepsilon_1|, \ldots, |\varepsilon_n|).$$

This table is correct up to the last printed digit.

Weighted least squares. With $\rho_1(\varepsilon) = \varepsilon^2$, the weight of the four outlying observations is gradually decreased to go smoothly from size 21 to size 17. This procedure implies prior identification of the outliers. In Table 2, Fit 1 is the ordinary least squares on size 21 whereas Fit 2 corresponds with size 17.

Least ν-th power. Selecting the form

$$\rho_1(\varepsilon) = |\varepsilon|^\nu$$

with ν between 2 and 1, a range of solutions with more or less incidence of the outlying observations is obtained by the algorithm of Rey (1975a) or, at less expense, by the method of next section. The result for ν = 1.2 is reported as Fit 3.

Huber's method. As seen previously, we select

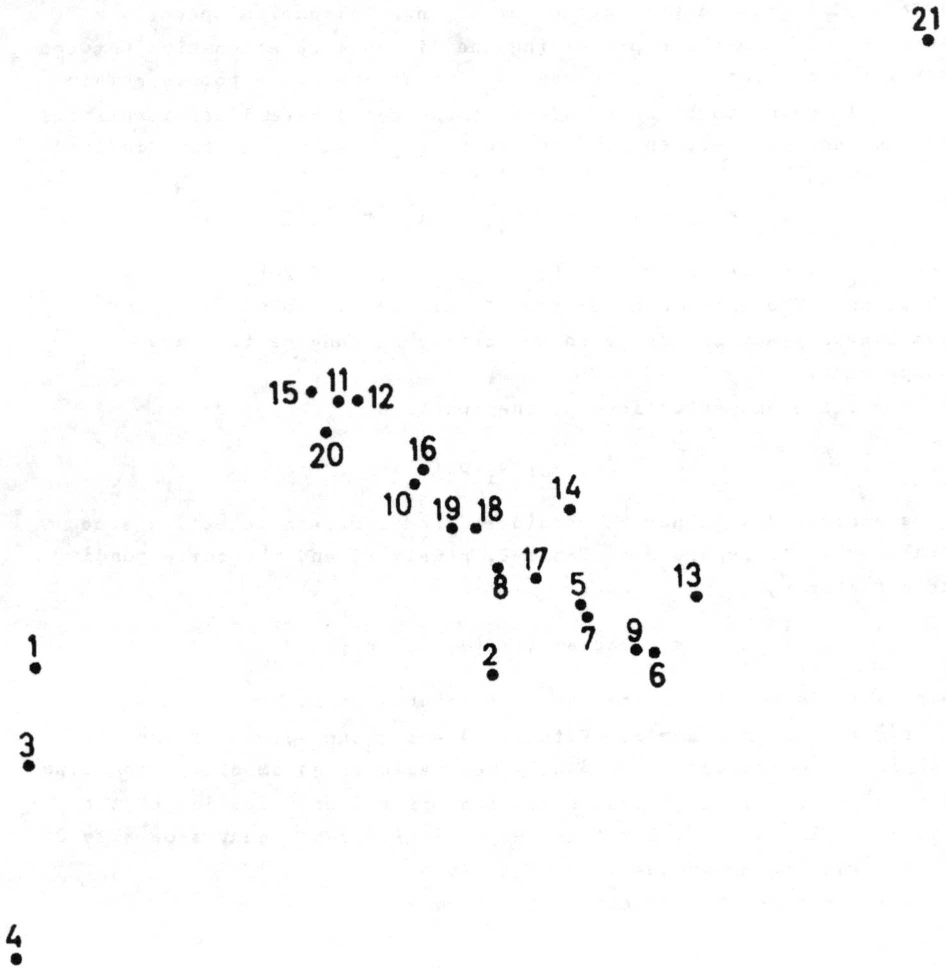

FIG. 1

$$\rho_1(\varepsilon) = \varepsilon^2, \qquad \text{if } \varepsilon > ks,$$
$$= ks(2\,|\varepsilon| - ks), \qquad \text{if } \varepsilon > ks.$$

The result for k = 1 is reported as Fit 4. Let us note that the method proposed by Huber is the only admissible method among the four we are considering.

Method of Andrews. The selected structure is

$$\rho_1(\varepsilon) = 1 - \cos[\varepsilon/(cs)], \qquad \text{if } |\varepsilon| \leq \pi cs$$
$$= 2, \qquad \text{if } |\varepsilon| > \pi cs.$$

A result for c = 1.5 is reported as Fit 5. It does not correspond strictly with Andrews' result (1974, Table 5, last but 1 column) due to large inaccuracies in his computation. The claim that the result is the same with as without the four outlying observations is nonsense, seeing that the scale factor s on size 21 and on size 17 differ significantly. In fact, the size 21 result is a nearly correct solution whereas the size 17 result is not defensible. Possibly the most important weakness of this method is that it may produce aberrant results in an unnoticed way. There are discontinuities in $\underline{\theta}_1$ as a function of the parameter c and there may be several solutions for a given c value. Fit 6, Fit 7 and Fit 8 are three solutions obtained with c = 1.8, a value intermediate between the two Andrews recommends, 1.5 and 2.1.

Fit	$\theta_{1,1}$	$\theta_{2,1}$	$\theta_{3,1}$	$\theta_{4,1}$	s	
1	-39.920	.71564	1.2953	-.15212	1.9175	L.sq., size=21
2	-37.652	.79769	.57734	-.06706	1.0579	L.sq., size=17
3	-38.805	.82643	.64760	-.08577	1.2194	Lν, ν = 1.2
4	-38.158	.83800	.66290	-.10631	1.1330	Huber, k = 1
5	-37.132	.81829	.51952	-.07255	.96533	Andrews, c=1.5
6	-37.334	.81018	.54199	-.07037	.99926	Andrews, c=1.8
7	-41.551	.93911	.58026	-.11295	1.4385	Andrews, c=1.8
8	-41.990	.93352	.61946	-.11278	1.5710	Andrews, c=1.8

Table 2

Each method produces a different $\underline{\theta}_1$ estimate, but there are trends which become apparent while comparing the respective fit residuals, ε_i.

All methods tend to reject outliers in a similar way, but differently take into account the non outlying observations.

The scale parameter s is strongly dependent upon the estimated $\underline{\theta}_1$ - see last column of Table 2 - and fixed point solutions are peremptory.

The method proposed by Andrews is totally deceiving. The method proposed by Huber wins the competition and is closely followed by the least ν-th power method. The latter is more expensive than the former in computation time.

There remains an important conceptual difficulty with regard to understanding exactly what has been estimated.

4.5. Solution of fixed point and non-linear equations

In the last section, we have seen that the methods involved in the computation of MM-estimators can be, and frequently are, rather time-consuming when the relatively obvious methods of relaxation or of simultaneous solution are attempted. We have also observed that one-step method could be inaccurate and, possibly, could dissimulate divergence of the otherwise-iterative process. Even with M-estimators difficulties can be encountered (see section 4.3.2). This section will be centered on fixed point considerations and application of the continuation method to solve non-linear equations.

Basically all converging iterative methods can be seen as fixed point computations. In our context of MM-estimation, given an approximate solution $(\theta_1^*,\dots,\theta_g^*)$ we work out through some arithmetic rule R'(.) an "improved" approximate solution.

$$(\theta_1^{**},\dots,\theta_g^{**}) = R'(\theta_1^*,\dots,\theta_g^*)$$

and repeat the process until stationarity is obtained, that is to say until

$$(\theta_1^*,\dots,\theta_g^*) = R'(\theta_1^*,\dots,\theta_g^*).$$

Then, the solution is

$$(\theta_1,\dots,\theta_g) = R'(\theta_1^*,\dots,\theta_g^*).$$

The solution of the general fixed point problem still presents many difficulties. We will first consider the situation where we know for sure there is a single solution; then we will take care of multiple solutions.

We note that it is equally difficult to apply the Brouwer theorem as to define some contracting mapping through a Lipschitz constant due to the impossibility of delineating a compact subset in the space of the parameters, except in fairly trivial situations - See Henrici (1974, section 6.12) in this respect - Thus, we will rather devote our attention to a general computational algorithm investigated by Scarf (1973) and further refined in Kellogg et al. (1976). A few complementary aspects and practical considerations are reported by Todd (1976).

The algorithm is based on the "continuation method" and consists in following "the" solution when some parameter varies. Precisely, assume we want the single solution of

$$\theta = R(\theta)$$

with

$$\theta = (\theta'_1, \ldots, \theta'_g)'$$

and assume we know the solution for another rule $R_0(.)$

$$\theta_0 = R_0(\theta_0),$$

than θ is the solution for $\lambda = 1$ of

$$\theta = \lambda\, R(\theta) + (1 - \lambda)\, R_0(\theta).$$

The method consists in following the solution from $\lambda = 0$, where it is $\theta = \theta_0$, up to $\lambda = 1$ where it is the fixed point solution desired. The efficiency of the method is the greater, the smoother is the implicit function $\theta(\lambda)$ with respect to the "variable" λ.

The continuation of the solution from $\lambda = 0$ up to $\lambda = 1$ presents no serious problem, whatever the method is, as long as the (matricial) expression

$$B = 1 - \lambda(\partial/\partial\theta)\, R(\theta) - (1 - \lambda)\, (\partial/\partial\theta)\, R_0(\theta)$$

remains positive definite. Predictor-corrector algorithms as well as involved analytical treatment can be proposed. We favor the former seeing their good numerical stability and they have been used in the illustration of section 4.4.4. When the above expression becomes singular that indicates that $\theta(\lambda)$ is not differentiable for some $\lambda \in [0,1]$ or, in other terms, that the starting set θ_0 was not appropriate. What is happening ?

More involved analysis permits to see that starting a continuation with some θ_0 may lead to a termination at some λ_1 ($\lambda_1 < 1$), or to a discontinuity or, possibly, to an explosion of the continuation path into several paths. Further, with the help of a partition of the parameter space, it is possible to count the number of going-out paths for one going-in path entering a given part. A good account of the relevant theory can be found in Amann (1976, chapter 3); it is based on the Leray-Schauder degree for compact vector fields defined on the closure of open subsets of some Banach space. As far as we are here concerned, we retain the attention on a very useful property of the continuation method : assume that $R(\theta)$ and $R_0(\theta)$ are continuous one-to-one mappings of θ in the space of θ, further assume that $\theta_0 = \theta(0) = \theta(1)$, then $\theta(\lambda) = \theta_0$ for λ in $[0,1]$; moreover when θ_0 is in some neighborhood of $\theta(1)$, $\theta(\lambda)$ is continuous with respect to λ. This property leads to very fast computations when approximate fixed point solutions are known.

We now specialize the above discussion to the evaluation of MM-estimators. First let us split the expression

$$\theta = R(\theta)$$

in its components; we want a fixed point solution of

$$(\theta'_1,\ldots,\theta'_g) = (R_1(\theta)',\ldots,R_g(\theta)')$$

where we must select $R_i(\theta)$ according to the estimation theory, we select

$$R_i(\theta) = \theta_i - \int \psi_i(x,\theta_1,\ldots,\theta_g)\, f(x)\, dx.$$

The choice of the rule $R_0(.)$ may be fairly trivial, say we select

$$R_{0i}(\theta) = \theta_{0i},$$

where θ_{0i} is a constant of the right magnitude or, possibly, a less robust estimation of θ_i. With these definitions, the above factor B takes the form

$$B = (1 - \lambda)\, 1 + \lambda A$$

where A is the matrix of block-components

$$A_{ij} = \int \phi_{ij} (x,\theta_1,\ldots,\theta_g) \; f(x) \; dx$$

encountered at section 4.2. When B is positive definite and when θ^* is an approximation of $\theta(\lambda)$ - e.g., obtained by a predictor formula - an improved approximation is given by the Newton-Raphson

$$\theta(\lambda) = \theta^* + B^{-1} [\lambda \; R(\theta^*) + (1-\lambda) \; R_0(\theta^*) - \theta^*].$$

The procedure as it has been described generally converges fairly rapidly but from time to time hazards have been met with, associated with serious difficulty to achieve the continuation from $\lambda = 0$, up to $\lambda = 1$. For instance in the computations of section 4.4.4 with the regression method of Andrews, it has appared that $\theta = \theta(c)$ was a continuous function of the parameter c from $c = \infty$ down approximately to $c = 2.35$ and all computations were problem-free. Difficulties were met with the value $c = 2.3$; starting with the predicted set

$$\theta_0(2.3) = 2 \; \theta(2.35) - \theta(2.4),$$

it has been possible to realize the continuation only very slowly. This has produced $\theta(2.3)$. With this result as starting set

$$\theta_0(2.25) = \theta(2.3),$$

a fast computation has given the estimator $\theta(2.25)$ for $c = 2.25$. Then we have tried to again estimate the result corresponding with $c = 2.35$. With the initialization

$$\theta_0(2.35) = 2 \; \theta(2.3) - \theta(2.25),$$

a very rapid computation has come out with an estimate θ close to $\theta_0(2.35)$ but totally different from the already met $\theta(2.35)$. A second fixed point solution had been found for $c = 2.35$.

To conclude this section it may be suitable to indicate that the present fixed point mathematics is equivalent to the simultaneous solution of section 4.4.2, except that substitution of the function

$$\rho_i(x,\theta_1,\ldots,\theta_g)$$

must be performed by

$$\lambda\, \rho_i(x,\theta_1,\ldots,\theta_g) + (1 - \lambda)\rho_{0i}\, (x,\theta_1,\ldots,\theta_g)$$

with

$$\rho_{0i}(x,\theta_1,\ldots,\theta_g) = (1/2)\, (\theta_i - \theta_{0i})'\, (\theta_i - \theta_{0i}).$$

Possibly the greatest advantage of the fixed point approach is to provide a larger insight retaining the attention on the multiplicity aspects as well as on the continuation method. The generality of the fixed point concept is well illustrated by the series of papers edited by Swaminathan (1976) and by Karamardian (1977) concerning mathematical theories as well as algorithmic features.

5. OPEN AVENUES

We present in this chapter a few considerations which either seem worthy of further research or appear to us rather weakly justified.

5.1. Estimators seen as functional of distributions

Section 2.4 is at the root of most derivations presented in this work however, and the fact must be stressed, it has been hardly possible to state precisely under what conditions a functional T(.), evaluated on distribution g, can be expanded with respect to distribution f according to

$$T(g) = T(f) + \int \psi(x)\, g(x)\, dx + \frac{1}{2} \iint \psi(x,y)\, g(x)\, g(y)\, dx\, dy + \ldots$$

This state of affair is unsatisfactory although the high plausibility of the expansion has been demonstrated for a large class of functionals and distributions. As already noted the attempts by many groups of experts to state precise conditions have failed. This failure can partly be attributed to the complexity of the problem but, possibly, also to the shortage of motivated analysts with top-qualification in topologies. - Furthermore, it is not clear whether this expansion is really required.

What we really need is an expansion which can be truncated to the level of its first few terms and, possibly, this is verified even for some diverging infinite expansions. Moreover, the influence function and the jackknife theory could have been presented with a completely heuristic basis, as they have previously; they have their values per se and, in case of doubt, it may be possible to assess by simulation what has been derived.

With regard to the influence function, Mallows, in an experimental investigation (1975), fully supports the definition based on the first von Mises derivative, even for moderate size samples. His criteria involve a few "natural" requirements (consistency, being a statistic, everywhere defined,...). Mallows also makes use of the above expansion to introduce a second-order influence function; it happens to be the integrand $\psi(x,y)$ of the double integral and is defined as the influence function of the (first-order) influence function. This may be a way to tackle (linearly) correlated data.

5.2. Sample distribution of estimators

Relatively few pieces of information are available regarding the sample distributions of robust estimators. We have already given with the jackknife method ways of assessing the possible bias, the variance and the symmetry (by third central moment) of their distributions. Furthermore, we may safely conjecture that the tendency toward the normal distribution is rapid whenever the influence function is bounded.

This state of affair is unsatisfactory although the question has the utmost practical interest. Attempts by Huber (1968) to set confidence limits have supplied some more insight on the relations between the distribution of the robust location estimator and the distribution of the sample; but these attempts have essentially demonstrated the difficulty of the problem. Subsequently, he has proposed a few conjectures on studentization of robust estimates (1970); they appear to us very reasonable. In the same line of tendency to normality, the paper on the computation of the sample distribution by Hampel (1973b) is noteworthy; we excerpt :

> "... the third and perhaps most important point seems to be entirely new. It concerns not the question where to expand, but what to expand. Most papers consider the cumulative distribution F_n, some the density f_n, but neither approach leads to very simple expressions. It shall be argued that the most natural and simple quantity to sudy is the derivative of the logarithm of the density, f'_n/f_n, and this for several reasons : ..."

whether his approach is practical remains to be demonstrated.

It may be seen that we feel entangled in a vicious circle : we make use of robust estimators because we do not know precisely the distribution of the sample at hand, and we should know how a sample departs from a given model in order to state the distribution of the estimator. We are afraid it is fighting for a wrong issue to try to derive theoretical distributions for robust estimators. We are afraid the only possibility is inference from the sample itself, in spite of the evident limitations of this method. We feel that for small sample sizes it is not wise to assume any strict model whereas, for moderate or large sample sizes, models are not needed seeing the tendency to normality.

But we may also question whether we are able to estimate the mean and the scatter to apply in studentization. The answer is positive, even for small sizes, with the help of the jackknife method. We illustrate.

For 5000 replicates of size $n = 11$, $(u_1,\ldots,u_{11})$, drawn from the negative exponential distribution

$$f(u) = \exp[-(u - a)], \quad \text{if } u \geq a,$$
$$= 0, \quad \text{if } u < a,$$

we have estimated the location parameter a by the method of Huber, i.e. we have estimated θ such that

$$\sum \rho(u_i - \theta) = \text{min for } \theta$$

with

$$\rho(\varepsilon) = \varepsilon^2, \quad \text{if } |\varepsilon| \leq k$$
$$= k(2|\varepsilon| - k), \quad \text{if } |\varepsilon| \geq k.$$

When we note θ_0 the asymptotic value of θ, i.e. θ_0 satisfies

$$\int \rho(u - \theta_0)\, f(u)\, du = \text{min for } \theta_0,$$

we have that $\theta - \theta_0$ is a consistent estimator of parameter a. Moreover, the variance of θ is approximately

$$\sigma_\theta^2 = \text{var}(\theta) = \frac{1}{n-1} V$$

where V is the asymptotic variance according to the jackknife theory.

Through analytical derivations we have obtained the expressions giving θ_0 and σ_θ^2 as functions of the parameter k. Thus we are in a position to compare the theoretical and experimental results for a relatively small sample size.

In Table 3, we report the results for two specific values, $k = .1$ and $k = 2$. θ_0 and σ_θ are the theoretical values to be compared with $\overline{\theta}$ and s_θ, the experimental mean and standard deviation observed with the 5000 replicates. Except for a neglegible bias with $k = .1$, we note a perfect agreement between theory and simulation notwithstanding the small size of the sample, $n = 11$.

	k = .1	k = 2
θ_0	.6948	.9475
$\overline{\theta}$	.7370	.9530
σ_θ	.3055	.2768
s_θ	.2993	.2632
$\overline{\theta} - \theta_0$	.0422	.0055
$(\overline{\theta} - \theta_0)/(\sigma_\theta/\sqrt{5000})$	9.768	1.405

Table 3

We may also compare the experimental distributions for these two values of k. Figure 2 is a display of the two probability density functions (pdf) against the relative coordinate $(\theta - \overline{\theta})/\sigma_\theta$. We note that the positive tails are similar while the negative tails differ from one another in a way which could be expected. We refrain from further commenting due to the fact that the figure is derived from 5000 replicates, a relatively small number in this regard.

5.3. Adaptive estimators

Seeing the inadequacy of some estimators when applied to inappropriate distributions, many authors have tried to design sets of estimators, each member of the sets being optimal for a specific class of distributions. Then, which estimator to select for a given sample is decided by means either of tests or by setting weights. To illustrate, consider the location estimator

$$\theta = w(\text{mean}) + (1 - w)\ (\text{median}).$$

Based on a sample $(x_1,\ldots,x_n)$, a test may indicate the high likelihood of a heavy-tailed distribution (e.g., see Smith, 1975). The first term of the above alternative consists in setting w to zero when the test favors the heavy-tail hypothesis, and to one otherwise; the second term of the alternative would be to define w as a monotonously increasing function of some test statistics. - A good account of the advantages and deficiencies of adaptive methods is presented by Hogg (1974), with

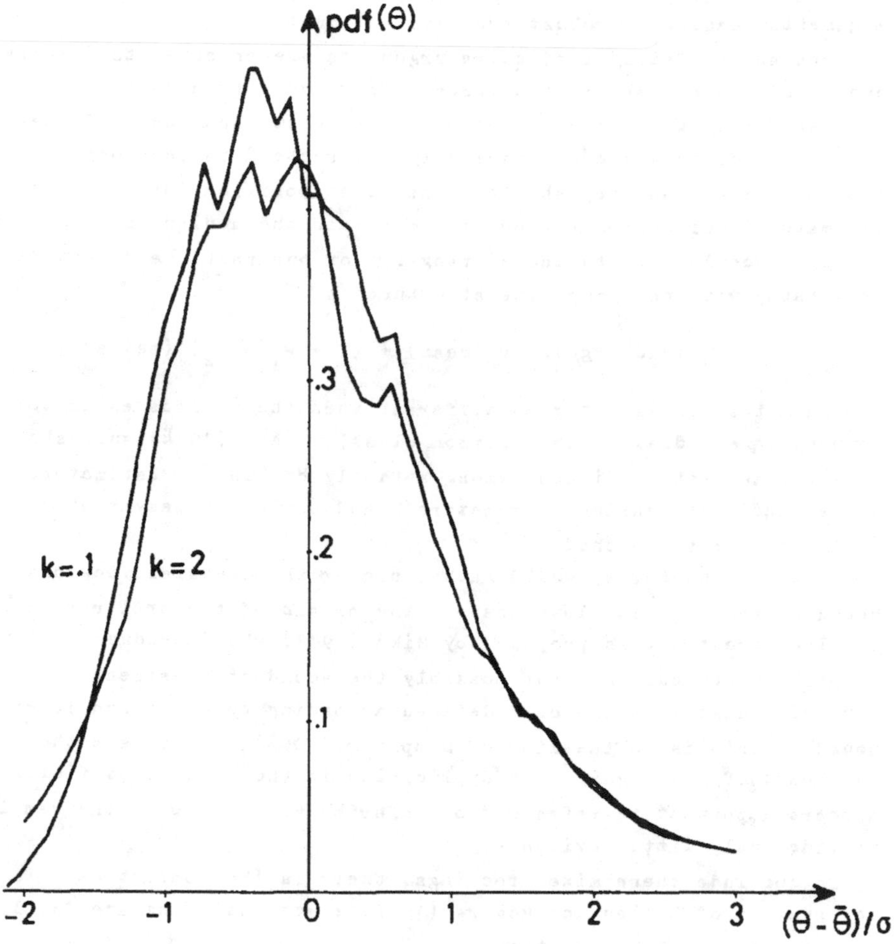

FIG. 2

particular emphasis on the historical background as well as the robust aspects. The paper of Takeuchi (1975) has possibly a broader scope although being less informative; it is essentially a discussion of the arguments leading to robust adaptive estimators.

Frequently, fairly subjective arguments are proposed to justify the choice of the estimator structures. They range from saying that most samples are drawn from mixtures of normal distributions to declaring that man-made data are unconsciously "corrected" to be closer from some typical value than they should. The first position favors a location estimator intermediate between the mean and the median, but the second viewpoint could lead to the midrange. For our part, we do not feel very happy with the composite structure

$$\theta = w_1 \text{ (midrange)} + w_2\text{(mean)} + (1 - w_1 - w_2) \text{ (median)}.$$

The situation is not much different when the various estimator components are drawn from a common class. Yohai (1974) suggests to select the "best" estimator among a family of Huber's estimators corresponding to various parameters k and proves the asymptotic optimality of the method.

In this context, we would rather prefer the bayesian viewpoint because, then, it is clear what is the origin of the arbitrariness. An involved treatment is proposed by Miké (1973) who introduces a family of prior distributions; but possibly the soundest treatment would be to use prior distributions only defined according to upper and lower bounds. This is in the line of Dempster (1968), but this author has not heard of any significant application in the robustness field. It appears important to refrain from infering too much when the sample provides only little evidence.

To conclude these mixed feelings, there is the comforting observation of Relles and Rogers (1977) : statisticians are fairly robust estimators of location.

5.4. Recursive estimators

Mainly in time-series analysis, it is peremptory to have expressions which permit to work out an estimator on a sample of size (n + 1), when the estimator on size n as well as some summary information are given. Makhoul (1975) surveys the various methods relevant to forecasting with linear models and they appear non-robust.

The design of robust methods is, so far, unsatisfactory. Some theory for recursive estimators of type M has been developped by Nevel'son (1975); it presents similarities with the Robbins-Monro algorithm and is relative to one-step M-estimators. Whether the finite sample properties are satisfactory remains an open question.

With regard to filtering, Tollet (1976) proposes a Kalman-Bucy approach with a gain factor selected in order to bound the sensitivity to gross errors or outliers. Occasional outliers have a weak influence, but that influence is not strictly bounded in order to permit tracking in presence of a large offset. A more accessible paper by Masreliez and Martin (1977) is partly similar. More or less the same strategy had been proposed by Rey (1974) with a recursive estimation of the "median" of serial data. That work also provides a robust estimation of the scale analogous with the proposal of Kersten and Kurz (1976), based on the paper of Evans et al. (1976).

In general, the analysis of time series with recursive robust estimators is based, at least in part, on heuristic approaches derived from the Robbins-Monro algorithm. However, it is rare to obtain satisfactory convergence. The reasons are, on the one hand, that lack of stationarity must be coped with and that, on the other hand, it is hardly possible to build any appropriate compact representation of the past information. These difficulties are encountered even when the non-linearities involved in the analytical treatment are moderate as can be seen in Rey (1975a).

It does not seem that robust estimation is considered in the investigation of point processes, except occasionally. Gaver and Hoel (1970) have compared several estimates of a reliability factor in a Poisson process with emphasis placed on bias, variance and sensitivity to Poisson hypothesis. They conclude that an estimate derived from the jackknife theory is optimal in most respects.

In time-series as well as in point process analyses, the limited results can be attributed to difficulty to assess what are the key assumptions of the models and to the difficulty of evaluating the sensitivity to these assumptions. Among them, the independency assumption is nearly unmanageable; it is frequently opposed to linear correlation, but this is only one of the possible alternate hypotheses. This question is approached with the help of an influence function by Devlin et al. (1975) and by Mallows (1975); the results are too complicate to be very promising.

5.5. Other views in robustness

In section 4, we have devoted great attention to the solution of "linear" regression problems without questionning the model validity. This question is directly addressed by Huber (1975) with a minimax approach partly based on the alternative, either linear or quadratic; this paper has been criticized and slightly augmented by Marcus and Sacks (1977). Even in the simplest cases, it seems that a tremendous amount of work is needed to cope with hidden non-linearities. More general polynomial structures are frequently assumed with eventual deletion of the "abusively" introduced terms. This is particularly discussed by Box and Draper (1975) as well as by Mead and Pike (1975) with regard to the robustness of the approach. The techniques for deleting variables are reviewed by Hocking (1976) and an interesting situation with discontinuous data is presented by Dyke (1974). We are here facing the general problem of the best selection of variables; the branch and bound method should not be disregarded when it is cleverly implemented (e.g., Pearsall, 1977) in spite of its shortcomings put to light by Cover and van Campenhout (1977). Non-polynomial structures can also be concerned, when due care is taken of the possible non-linearities; this may be done by ridge regression (see Hoerl and Kennard (1976) for a recent paper on their method) or the non-linearities can be partly attributed to errors in the variables, as reviewed by Florens et al. (1974). What is the most appropriate attitude is very much dependent upon the problem at hand and can hardly be discussed in the present text. But we retain the attention on the fact that what is precisely estimated is not clear. Most methods can better be seen as approximation methods than as statistical procedures.

To conclude let us recall that robust estimation in data analysis is, for us, an essential counterpart to the more classical methods. Robust methods help in validating computation results by providing a reliable comparison basis. The need for these methods is particularly evident while processing multivariate data, as advocated by Gnanadesikan and Wilk (1969) and by Gnanadesikan and Kettenring (1972). In the multivariate domain, the identification of outliers may present formidable difficulties (e.g., Rohlf, 1975) and robust methods not requiring this identification are welcome. This is illustrated by Figures 3 and 4, the former is a classical least squares regression whilst the latter is a regression obtained by Huber's method as described at section 4.4.3. Which is the best regression is a matter of opinion, but the second may be preferred and (this is the most important) the difference between the two results is such that the sample is worthy of further attention.

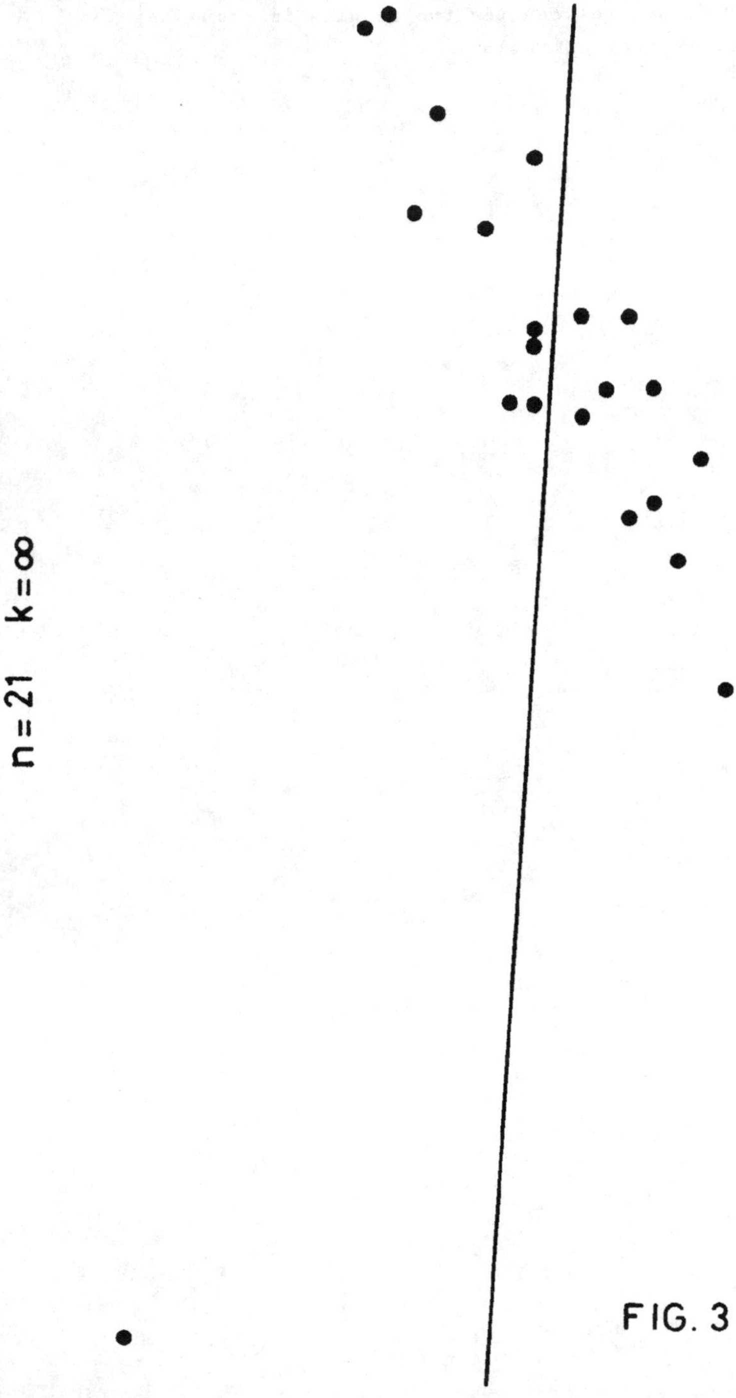

FIG. 3

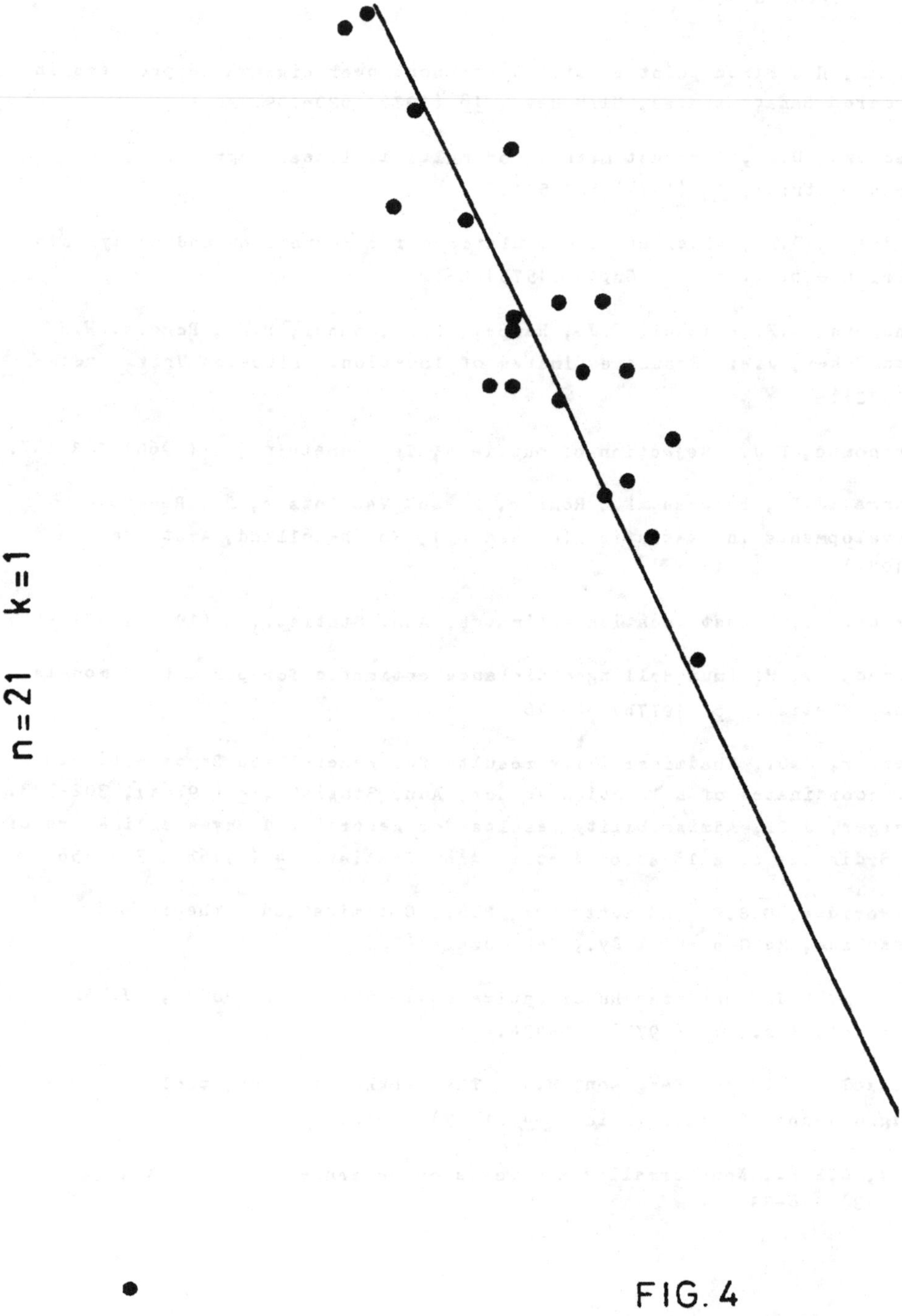

FIG. 4

6. REFERENCES

Amann, H., Fixed point equations and nonlinear eigenvalue problems in ordered Banach spaces, SIAM Rev., 18 (1976) 620-709.

Andrews, D.F., A robust method for multiple linear regression, Technometrics, 16 (1974) 523-531.

Andrews, D.F., Alternative calculations for regression and analysis of variance problems, In Gupta (1975) 2-7.

Andrews, D.F., Bickel, P.J., Hampel, F.R., Huber, P.J., Rogers, W.H. and Tukey, J.W., Robust estimates of location, Princeton Univ. Press (1972).

Anscombe, F.J., Rejection of outliers, Technometrics, 2 (1960) 123-147.

Barra, J.R., Brodeau, F., Romier, G. and Van Cutsem, B., Recent developments in statistics (edited by), North-Holland, Amsterdam (1977).

Beran, R., Robust location estimates, Ann. Statist., 5 (1977a) 431-444.

Beran, R., Minimum Hellinger distance estimates for parametric models, Ann. Statist., 5 (1977b) 445-463.

Berger, J.O., Inadmissibility results for generalized Bayes estimators of coordinates of a location vector, Ann. Statist., 4 (1976a), 302-333.

Berger, J.O., Admissibility results for generalized Bayes estimators of coordinates of a location vector, Ann. Statist., 4 (1976b), 334-356.

Beveridge, G.S.G. and Schechter, R.S., Optimization : theory and practice, Mc Graw-Hill Cy., New York (1970).

Bickel, P.J., One-step Huber estimates in the linear model, J. Am. Statist. Ass., 70 (1975) 428-434.

Bissel, A.F., and Ferguson, R.A., The jackknife : Toy, tool or two edged weapon ? Statistician, 24 (1975) 79-100.

Box, G.E.P., Non-normality and tests on variances, Biometrika, 40 (1953) 318-335.

Box, G.E.P. and Draper, N.R., Robust designs, Biometrika, 62 (1975) 347-352.

Boyd, D.W., The power method for Lp norms, Linear Algebra Appl., 9 (1974) 95-101.

Brownlee, K.A., Statistical theory and methodology in science and engineering, Wiley, New York, 2nd ed. (1965).

Cargo, G.T. and Shisha, O., Least p-th powers of deviations, J. Approximation Thy, 15 (1975) 335-355.

Chen, C.H., On information and distance measures, error bounds and feature selections, Information Sciences, 10 (1976) 159-174.

Coleman, D., Holland, P., Kaden, N., Klema, V. and Peters, S.C., A system of subroutines for iteratively reweighted least squares computations, manuscript, Mass. Inst. Techn., (1977).

Collins, J.R., Robust estimation of a location parameter in the presence of asymmetry, Ann. Statist., 4 (1976) 68-85.

Cover, T.M. and van Campenhout, J.M., On the possible orderings in the measurement selection problem, IEEE Trans. Syst. Man Cyb., SMC-7 (1977) 657-661.

Daniel, C. and Wood, F.S., Fitting equations to data, Wiley, New York (1971).

Dempster, A.P., Estimation in multivariate analysis, In Krishnaiah (1966) 315-334.

Dempster, A.P., A generalization of bayesian inference, J. Roy. Statist. Soc., B-30 (1968) 205-232.

Dempster, A.P., A subjectivist look at robustness, In I.S.I., 1 (1975) 349-374.

Dempster, A.P., Examples relevant to the robustness of applied inferences, In Gupta and Moore (1977) 121-138.

Denby, L. and Mallows, C.L., Two diagnostic displays for robust regression analysis. Technometrics, 19 (1977) 1-13.

Devlin, S.J., Gnanadesikan, R. and Kettenring, J.R., Robust estimation and outlier detection with correlation coefficients, Biometrika, 62 (1975) 531-545.

Draper, N.R. and Smith, H., Applied regression analysis, Wiley, New York (1966).

Dutter, R., Algorithms for the Huber estimator of multiple regression, Computing, 18 (1977) 167-176.

Dyke, G.V., Designs to minimize loss of information in polynomial regression, Appl. Statist., 23 (1974) 295-299.

Eisenhart, C., The development of the concept of the best mean of a set of measurements from antiquity to the present day, 1971 Am. Statist. Ass. Presidential Address, Audience notes (1971).

Ekblom, H., Calculation of linear best Lp approximations. BIT (Nordisk Tidskr. Informationsbehandling), 13 (1973) 292-300.

Ekblom, H., Lp-methods for robust regression, BIT (Nordisk Tidskr. Informationsbehandling), 14 (1974) 22-32.

Evans, J.G., Kersten, P. and Kurz, L., Robust recursive estimation with applications, Information Sciences, 11 (1976) 69-92.

Feller, W., An introduction to probability theory and its applications, Wiley, New York, Vol. 2 (1966).

Ferguson, R.A., Fryer, J.G. and Mc Whinney, I.A., On the estimation of a truncated normal distribution, In I.S.I., 3 (1975) 259-263.

Filippova, A.A., Mises' theorem of the asymptotic behavior of functionals of empirical distribution functions and its statistical applications, Theor. Probab. Appl., 7 (1962) 24-57.

Fine, T.L., Theories of probability; an examination of foundations, Academic Press, New York (1973).

Fletcher, R., Grant, J.A. and Hebden, M.D., The continuity and differentiability of the parameters of best linear Lp approximations, J. Approximation Thy, 10 (1974) 69-73.

Fletcher, R. and Powell, M.J.D., A rapidly convergent descent method for minimization, Computer J., 6 (1963) 163-168.

Florens, J.P., Mouchart, M. and Richard J.F., Bayesian inference in error-in-variables models, J. Multivariate Anal., 4 (1974) 419-452.

Forsythe, A.B., Robust estimation of straight line regression coefficients by minimizing p-th power deviations, Technometrics, 14 (1972) 159-166.

Garel, B., Détection des valeurs aberrantes dans un échantillon gaussien multidimensionnel, Thèse, Univ. Grenoble (1976).

Gaver, D.P. and Hoel. D.G., Comparison of certain small-sample Poisson probability estimates, Technometrics, 12 (1970) 835-850.

Gentleman, W.M., Robust estimation of multivariate location by minimizing p-th power deviations, Ph. D. Dissertation, Princeton University, and Memorandum MM 65-1215-16, Bell Tel. Labs. (1965).

Gnanadesikan, R. and Kettenring, J.R., Robust estimates, residuals and outlier detection with multiresponse data, Biometrics, 28 (1972) 81-124.

Gnanadesikan, R. and Wilk, M.B., Data analytic methods in multivariate statistical analysis, In Krishnaiah (1969) 593-638.

Gray, H.L. and Schucany, W.R., The generalized jackknife statistic, Marcel Dekker, New York (1972).

Gray, H.L., Schucany, W.R. and Watkins, T.A., On the generalized jackknife and its relation to statistical differentials, Biometrika, 62 (1975) 637-642.

Gray, H.L., Schucany, W.R. and Woodward, W.A., Best estimates of functions of the parameters of the gaussian and the gamma distributions, IEEE Trans. Reliability, R-25 (1976) 95-99.

Gross, A.M., Confidence intervals for bisquare regression estimates, J. Am. Statis. Ass., 72 (1977) 341-354.

Gross, A.M. and Tukey, J.W., The estimators of the Princeton Robustness Study, Princeton Univ., Techn. Rept. 38-2 (1973).

Gupta, R.P., Applied Statistics (edited by), North Holland, Amsterdam (1975).

Gupta, S.S. and Moore, D.S., Statistical decision theory and related topics (edited by), Academic Press, (1977).

Hampel, F.R., Contributions to the theory of robust estimation, Ph. D. Dissertation, Univ. California, Berkeley (1968).

Hampel, F.R., A general qualitative definitions of robustness, Ann. Math. Statist., 42 (1971) 1887-1896.

Hampel, F.R., Robust estimation : A condensed partial survey, Z. Wahrscheinlichkeitstheorie verw. Geb., 27 (1973) 87-104.

Hampel, F.R., Some small sample asymptotics, Proceed. Prague Symp. Asymptotic Statist., (1973b) 109-126.

Hampel, F.R., The influence curve and its role in robust estimation, J. Am. Statist. Ass., 69 (1974) 383-393.

Hampel. F.R., Beyond location parameters : Robust concepts and methods, In I.S.I. 1 (1975) 375-382.

Hampel, F.R., On the breakdown points of some rejection rules with mean, E.T.H. Zurich, Res. Rept 11 (1976).

Harter, H.L., The method of least squares and some alternatives, Int. Statist. Rev., 42 (1974) 147-174, 235-264, 282, 43 (1975) 1-44, 125-190, 273-278, 269-272, 44 (1976) 113-159.

Henrici, P., Applied and computational complex analysis, Vol. 1, Wiley, New York (1974).

Hill, R.W., Robust regression when there are outliers in the carriers, Ph. D. dissertation, Harvard University (1977).

Hille, E., Analytic function theory, Blaisdell, New York, Vol. 1 (1959).

Hocking, R.R., The analysis and selection of variables in linear regression, Biometrics, 32 (1976) 1-49.

Hoerl, A.E. and Kennard, R.W., Ridge regression : iterative estimation of the biasing parameter. Commun. Statist. - Theor. Meth., A5 (1976) 77-88.

Hogg, R.V., Adaptive robust procedures : A partial review and some suggestions for future applications and theory, J. Am. Statist. Ass., 69 (1974) 909-923.

Huber, P.J., Robust estimation of a location parameter, Ann. Math. Statist., 35 (1964) 73-101.

Huber, P.J., Robust confidence limits, Z. Wahrscheinlichkeitstheorie verw. Geb., 10 (1968) 269-270.

Huber, P.J., Théorie de l'inférence statistique robuste, Presses Univ, Montréal (1969).

Huber, P.J., Studentizing robust estimates, In Puri (1970) 453-463.

Huber, P.J., Robust statistics : A review, Ann. Math. Statist., 43 (1972) 1041-1067.

Huber, P.J., Robust regression: Asymptotics, conjectures and Monte Carlo, Ann. Statist., 1 (1973) 799-821.

Huber, P.J., Robustness and designs, In Srivastava (1975) 287-301.

Huber, P.J., Robust covariances, In Gupta and Moore (1977a) 165-191.

Huber, P.J., Robust methods of estimation of regression coefficients, Math. Operationsforsch. Statist., Ser. Statistics, 8 (1977b) 41-53.

Huber, P.J. and Dutter, R., Numerical solution of robust regression problems, COMPSTAT 1974, (1974) 165-172.

Hwang, S.Y., On monotonicity of Lp and lp norms, IEEE, Trans. Electroacoustic Speech Signal Proc., ASSP-23 (1975) 593-596.

I.S.I., Proceedings of the 40th Session, Warsaw - 1975, 4 books, Bull. Int. Stat. Inst., 46 (1975).

Jackson, D., Note on the median of a set of numbers, Bull. Ann. Math. Soc., 27 (1921) 160-164.

Jaeckel, L.A., Robust estimates of location : Symmetry and asymmetric contamination, Ann. Math. Statist., 42 (1971) 1020-1034.

Jaeckel, L.A., The infinitesimal jackknife, Bell Lab. Memorandum, MH-1215 (1972a).

Jaeckel, L.A., Estimating regression coefficients by minimizing the dispersion of the residuals, Ann. Math. Statist., 43 (1972b) 1449-1458.

Kanal, L., Patterns in pattern recognition, IEEE Trans. Information Theory, IT-20 (1974) 697-722.

Karamardian, S., Fixed points algorithms and applications (edited by), Academic Press, (1977).

Kellogg, R.B., Li, T.Y. and Yorke, J., A constructive proof of the Brouwer fixed point theorem and computational results, SIAM J. Numer. Anal., 13 (1976) 473-483.

Kendall, M.G. and Buckland, W.R., A dictionary of statistical terms (prepared for the International Statistical Institute). Oliver and Boyd, Edinburgh, 3rd ed. (1971).

Kersten, P. and Kurz, L., Robustized vector Robbins-Monro algorithm with applications to M-interval detection, Information sciences, 11 (1976) 121-140.

Komlós, J., Major P. and Tusnády, G., Weak convergence and embedding, In Revesz (1975) 149-165.

Krishnaiah, P.R., Multivariate analysis (edited by), Academic Press, 1 (1966).

Krishnaiah, P.R., Multivariate analysis (edited by), Academic Press, 2 (1969).

Lachenbruch, P.A., On expected probabilities of misclassification in discriminant analysis, necessary sample size, and a relation with the multiple correlation coefficient, Biometrics, 24 (1968) 823-834.

Lewis, J.T. and Shisha, O., Lp-convergence of monotone functions and their uniform convergence, J. Approximation Thy, 14 (1975) 281-284.

Makhoul, J., Linear prediction : a tutorial review, Proc. IEEE, 63 (1975) 693-708.

Mallows, C.L., On some topics in robustness, I.M.S. meeting, Rochester, May (1975).

Marcus, M.B. and Sacks, J., Robust designs for regression problems, In Gupta and Moore (1977) 245-268.

Maronna, R.A., Robust M-estimators of multivariate location and scatter, Ann. Statist., 4 (1976) 51-67.

Masreliez, C.J. and Martin, R.D., Robust bayesian estimation for the linear model and robustifying the Kalman filter, IEEE Trans. on Aut. Contr., AC-22 (1977) 361-371.

Mead, R. and Pike, D.J., A review of response surface methodology from a biometric viewpoint, Biometrics, 31 (1975) 803-851.

Merle, G. and Spath, H., Computational experiences with discrete Lp approximation, Computing, 12 (1974) 315-321.

Miké, V., Robust Pitman-type estimators of locations, Ann. Inst. Statist. Math., 25 (1973) 65-86.

Miller. R.G., The jackknife - a review, Biometrika, 61 (1974a) 1-15.

Miller, R.G., An unbalanced jackknife, Ann. Statist., 2 (1974b) 880-891.

Miller. R.G., Jackknifing censored data, Stanford University, Biostatistics Div., Techn. Rept., 14 (1975).

Mosteller, F., The jackknife, Rev. Int., Statist. Inst., 39 (1971) 363-368.

Munster, M., Théorie générale de la mesure et de l'intégration, Bull. Soc. Roy. Sc. Liège, 43 (1974) 526-567.

N.C.H.S., Annoted bibliography on robustness studies of statistical procedures, U.S. Dept. Health Educ. Welf., Publication (HSM) 72-1051 (1972).

Nevel'son, M.B., On the properties of the recursive estimates for a functional of an unknown distribution function, In Revesz (1975) 227-251.

Olkin, I., Contributions to probability and statistics (edited by), Stanford Univ. Press, (1960).

Pearsall, E.S., Best subset regression by branch and bound, Internal report, Wayne State University, (1977).

Prokhorov, Y.V., Convergence of random processes and limit theorems in probability theory, Theor. Probab. Appl., 1 (1956) 157-214.

Puri, M.L., Nonparametric techniques in statistical inferences (edited by), Cambridge Univ. Press, (1970).

Quenouille, M.H., Notes on bias in estimation, Biometrika, 43 (1956) 353-360.

Rao, C.R., Linear statistical inference and its applications - second edition, Wiley, New York (1973).

Reeds, J.A., On the definition of von Mises functionals, Ph. D. Dissertation and Res. Rept. S-44, Harvard University, (1976).

Relles, D.A. and Rogers, W.H., Statisticians are fairly robust estimators of location, J. Am. Statist. Ass., 72 (1977) 107-111.

Revesz, P., Limit theorems of probability theory (edited by) North-Holland, Amsterdam (1975).

Rey, W., Robust estimates of quantiles, location and scale in time series, Philips Res. Repts, 29 (1974) 67-92.

Rey, W., On least p-th power methods in multiple regressions and location estimations, BIT (Nordisk Tidskr. Informationsbehandling), 15 (1975a) 174-184.

Rey, W., Mean life estimation from censored samples, Biom. Praxim., 15 (1975b) 145-159.

Rey, W., M-estimators in robust regression, MBLE Research Laboratory, Brussels, R329 (1976).

Rey, W.J.J., M-estimators in robust regression - a case study, In Barra et al. (1977) 591-594.

Rey, W.J.J. and Martin, L.J., Estimation of hazard rate from samples, In I.S.I., 4 (1975) 238-240.

Rohlf, F.J., Generalization of the gap test for detection of multivariate outliers, Biometrics, 31 (1975) 93-101.

Ronner, A.E., P-norm estimators in a linear regression model, Dissertation, Groningen University, The Netherlands (1977).

Scarf, H., The computation of economic equilibria, Yale Univ. Press, New Haven and London (1973).

Scholz, F.W., A comparison of efficient location estimators, Ann. Statist., 2 (1974) 1323-1326.

Schucany, W.R., Gray, H.L. and Owen, D.B., On bias reduction in estimation, J. Am. Statist. Ass., 66 (1971) 524-533.

Schwartz, L., Analyse numérique; topologie générale et analyse fonctionnelle, Hermann, Paris (1970).

Sen, P.K., Some invariance principles relating to jackknifing and their role in sequential analysis, Ann. Statist., 5 (1977) 316-329.

Sharot, T., The generalized jackknife : finite samples and subsample sizes, J. Am. Statist. Ass., 71 (1976a) 451-454.

Sharot, T., Sharpening the jackknife, Biometrika, 63 (1976b) 315-321.

Smith, V.K., A simulation analysis of the power of several tests for detecting heavy-tailed distributions, J. Am. Statist. Ass., 70 (1975) 662-665.

Srivastava, J.N., A survey of statistical design and linear models (edited by), North-Holland, Amsterdam (1975).

Stein, C., A necessary and sufficient condition for admissibility, Ann. Math. Statist., 26 (1955) 518-522.

Stigler, S.M., Simon Newcomb, Percy Daniell and the history of robust estimation : 1885-1920, J. Am. Statist. Ass., 68 (1973) 872-879.

Swaminathan, S., Fixed point theory and its applications (edited by), Academic Press, (1976).

Takeuchi, K., A survey of robust estimation of location : models and procedures, especially in case of measurement of a physical quantity, In I.S.I., 1 (1975) 336-348.

Thorburn, D., Some asymptotic properties of jackknife statistics, Biometrika 63 (1976) 305-313.

Thucydides, History of the Peloponnesian war, 3 (428 B.C.) para 20, In Eisenhart (1971).

Todd, M.J., The computation of fixed points and applications, Lecture Notes in Econ. and Math. Syst., Springer Verlag, Berlin, 124 (1976).

Tollet, I.H., Robust forecasting for the linear model with emphasis on robustness toward occasional ouliers, IEEE Int. Conf. Cybernetics & Society, Washington (1976) 600-605.

Tukey, J.W., Bias and confidence in not-quite large samples (abstract), Ann. Math. Statist., 29 (1958) 614.

Tukey, J.W., A survey of sampling from contaminated distributions, In Olkin (1960) 448-485.

Vajda, S., Mathematical programming, Addison-Wesley Inc., Reading, Mas., (1961).

Von Mises, R., On the asymptotic distribution of differentiable statistical functions, Ann. Math. Statist., 18 (1947) 309-348.

Woodruff, R.S. and Causey, B.D., Computerized method for approximating the variance of a complicated estimate, J. Am. Statist. Ass., 71 (1976) 315-321.

Yale, C. and Forsythe, A.B., Winsorized regression, Technometrics, 18 (1976) 291-300.

Yohai, V.J., Robust estimation in the linear model, Ann. Statist., 2 (1974) 562-567.

Youden, W.J., Enduring values, Technometrics, 14 (1972) 1-11.

Ypelaar, A. and Velleman, P.F., The performance of robust regression procedures, Economic and social statistics technical reprint series, Cornell University, 877/007 (1977).

APPENDIX

The following pages are essentially devoted to the involved derivations which would unduely load the main text otherwise. The various sections have been ranked in alphabetical order to ease the research of specific developments, wherever they may be required. Many cross-references are made between the sections of the appendix. Then, the concerned entries are noted in italic type.

Consistent estimator

A vector estimate $\underline{t}_n$ is said to be consistent when it converges in probability, as the sample size n increases, to the parameter $\underline{\theta}$ of which it is an estimator - from Kendall and Buckland (1971). In other terms, $\underline{t}_n$ is consistent if, and only if,

$$\text{plim}\ \rho(\underline{t}_n, \underline{\theta}) = 0,$$

where $\rho(.,.)$ is the metric in the sample space Ω. This is equivalent to the vanishing of the *Prokhorov metric* $d(.,.)$ between f_n, the sample distribution of $\underline{t}_n$, and f, the Dirac distribution centered on $\underline{\theta}$.

$$\text{Consistency} \Leftrightarrow \lim d(f_n, f) = 0.$$

There may be interest in bounding the Prohkorov metric $d(f_n, f)$, with respect to moment of positive order α,

$$m_\alpha = E\{[\rho(\underline{t}_n, \underline{\theta})]^\alpha\}.$$

The Chebyshev-like inequality

$$\text{Prob}\{\rho(\underline{t}_n, \underline{\theta}) < \varepsilon\} \geqslant 1 - m_\alpha\, \varepsilon^{-\alpha}$$

leads to

$$d(f_n, f) \leqslant m_\alpha^{1/(\alpha+1)}.$$

Contaminated normal distributions

With the desire of assessing whether the normality assumption is important, Tukey (1960) has investigated the performances of several estimators on the normal and slightly non-normal distributions. We recall his findings and augment them with our own observations. He compares estimations of location and scale performed with a sample drawn from a purely normal distribution $N(\mu, \sigma^2)$ and with a sample drawn from the same normal distribution but contaminated by extraneous observations also normally distributed. His model of contaminated normal distribution has the form

$$(1 - \varepsilon)\ N(\mu, \sigma^2) + \varepsilon\ N(\mu_1, \sigma_1^2).$$

The symmetry is maintained for $\mu_1 = \mu$ and, for moderate ε, the contaminated model has shorter tails than the normal when $\sigma_1 < \sigma$ and longer when $\sigma_1 > \sigma$.

The findings may appear unexpected to some readers. It is well known that estimation of the normal distribution location by the median instead of the mean leads to an efficiency loss of 36 %, but a 10 % contamination is sufficient to have better estimation by the median. Concerning scale estimation the situation is rather exceptional; a contamination of one or two thousandths is sufficient to balance an efficiency loss of 12 % between the mean deviation and the standard deviation.

Precisely, in the following two tables, one will find the results obtained for some specific values of parameter ε. Table 4 is relative to the symmetric model

$$(1 - \varepsilon)\ N(0, 1) + \varepsilon\ N(0, 3^2),$$

whereas the asymmetric model

$$(1 - \varepsilon)\ N(0, 1) + \varepsilon\ N(2, 3^2)$$

has given table 5.

Location has been estimated by the mean and the median. Their respective asymptotic variances (var) constitute a natural basis for comparison. Scale has been assessed by the three following estimators : the standard deviation, the mean deviation to the mean and the median deviation to the median. A fourth estimator, the semi-interquartile range, has also been computed but has not been reported the results being the same as (or slightly poorer than) what we have with the median deviation to the median. Due to the lack of natural definition for the scale, it has appeared appropriate to compare the estimators on the basis of their respective coefficients of variation (c.v.), the ratio of their asymptotic standard deviations to their means.

The last lines of the tables provides a "discriminatory sample size" at level 0.95 which is the minimum size the sample must have to test whether observations are drawn from a strictly normal distribution or from a given contaminated distribution. We have supposed that a likelihood ratio test was performed to test H_0 against H_1, with

$$H_0 = N(\mu_\varepsilon, \sigma_\varepsilon^2)$$
$$H_1 = (1 - \varepsilon)\, N(0, 1) + \varepsilon\, N(0 \text{ or } 2, 3^2)$$

for given ε, and where μ_ε and σ_ε are the mean and the standard deviation relative to H_1. Numerical integrations have provided the size such that a sample drawn from H_0 be attributed to H_0 by the test with probability 0.95. It must be noted that our approach is very simple minded; more realistically we should take into account the estimation of the parameters and, then, the discriminatoy sample sizes would be larger.

Inspection of the tables reveals that the sensitivity to the distribution shape is very dependent upon the nature of the estimators. In particular, a slight long-tail contamination is sufficient to significantly impair the efficiencies of the mean and of the standard deviation. Moreover, quite large samples must be at disposal in order to test the optimality of these estimators.

Symmetric model

ε	0.0000	0.0018	0.0282	0.1006	0.2436	0.5235	0.8141
Mean							
-	0.0000	0.0000	0.0000	0.0000	0.0000	0.0000	0.0000
var	1.0000	1.0140	1.2252	*1.8047*	2.9488	5.1882	*7.5130*
Median							
-	0.0000	0.0000	0.0000	0.0000	0.0000	0.0000	0.0000
var	1.5708	1.5745	1.6315	*1.8047*	2.2389	3.7066	*7.5130*
Standard deviation							
-	1.0000	1.0070	1.1069	1.3434	1.7172	2.2778	2.7410
c.v.	0.7071	*0.7627*	*1.1725*	1.3540	*1.2317*	*0.9720*	0.7929
Mean deviation to the mean							
-	0.7979	0.8007	0.8428	0.9584	1.1866	1.6333	2.0970
c.v.	0.7555	*0.7627*	0.8514	0.9822	1.0461	*0.9720*	0.8417
Median deviation to the median							
-	0.6745	0.6754	0.6891	0.7297	0.8259	1.1089	1.6167
c.v.	1.1664	1.1668	*1.1725*	1.1899	*1.2317*	1.3435	1.3600
Discriminatory sample size							
0.95	-	7880	497	137	83	123	638

Table 4

Asymmetric model

ε	0.0000	0.0008	0.0115	0.0617	0.2228	0.6246	0.6878
Mean							
-	0.0000	0.0016	0.0230	0.1234	0.4456	1.2492	1.3756
var	1.0000	1.0097	1.1377	*1.7254*	3.4753	*6.9348*	7.3613
Median							
-	0.0000	0.0005	0.0072	0.0401	0.1657	0.7409	0.8986
var	1.5708	1.5727	1.5977	*1.7254*	2.2901	*6.9348*	8.7865
Standard deviation							
-	1.0000	1.0049	1.0666	1.3136	1.8682	2.6334	2.7132
c.v.	0.7071	*0.7611*	*1.1694*	1.5176	*1.2520*	0.8172	*0.7837*
Mean deviation to the mean							
-	0.7979	0.7996	0.8222	0.9301	1.2841	2.0487	2.1355
c.v.	0.7555	*0.7611*	0.8263	0.9973	1.0525	0.8077	*0.7837*
Median deviation to the median							
-	0.6745	0.6749	0.6810	0.7115	0.8368	1.4726	1.6093
c.v.	1.1664	1.1666	*1.1694*	1.1838	*1.2520*	1.3731	1.3080
Discriminatory sample size							
0.95	-	8760	769	121	51	119	164

Table 5

Distribution space

All probability functions, we are concerned with, are defined in a convex complete metric space, included in a Banach space. See Schwartz (1971, Ch. 11) and Hille (1959, Sec. 4.7).

For any three functions f, g and h belonging to this space E,

$$f,\ g,\ h \in E,$$

we define a metric by a distance function d(.,.) having the ordinary properties

$$\begin{aligned} d(f,g) &\geq 0 \\ d(f,g) &= 0 \Leftrightarrow f = g \\ d(f,g) &= d(g,f) \\ d(f,g) &\geq d(f,h) + d(h,g). \end{aligned}$$

The space is complete or, in other terms, the limit of any Cauchy sequence $\{f_n\}$ of E belongs to E, i.e.

$$\begin{aligned} &\{f_n\} \in E \\ &\lim\ \{f_n\} = f \Rightarrow f \in E. \end{aligned}$$

The space is convex, i.e.

$$\begin{aligned} &f,g \in E,\ t \in R,\ 0 \leq t \leq 1 \\ &h = tf + (1-t)g \Rightarrow h \in E. \end{aligned}$$

So far, we have not clarified what probability function we consider. With the usual terminology, they can be seen as being either frequency distributions or probability density distributions and they are related to some probability measure in the following way.

Let Ω be the whole sample space and $\omega \subset \Omega$ be some subset, then we associate the probability measure $F(\omega)$ to the probability function f by

$$F(\omega) = \int_{\omega} f(x)\ dx.$$

Whenever f(x) is not continuous for any $x \in \omega$, the integral notation must be understood in the sense of the distribution theory.

The sample space Ω can possibly be different for each distribution f, however we do not consider this possibility and we assume that it can be extended in order to be common to all distributions f and that it has a unit probability measure, i.e.,

$$F(\Omega) = 1.$$

This possible extension of the sample space Ω leads to see discrete distribution functions as a sum of Dirac functions rather than as smooth functions only defined on some discrete sample space. For metric construction purposes, we also require the sample space Ω to be metric.

It must be observed that we have as much as possible avoided the axiomatic theory of probability introduced by Kolmogorov, as related by Feller (1966, Chap. 4); this is because we have found it too constraining and abusively heavy to manipulate. A critical appraisal of the Kolmogorov setup has been reported by Fine (1973, Chap. 3) and justifies our attitude.

The selection of an appropriate metric is especially difficult. The only satisfactory proposal, to the best of our knowledge, is due to *Prokhorov* and may be seen as a multidimensional generalization of the Levy metric while $\Omega = \mathbb{R}^p$, the p-dimension real space. The *Prokhorov metric* permits to simultaneously include continuous and discrete probability functions in the distribution space. - Although we have required the possibility of considering simultaneously both types of distributions, we must indicate that several papers escape this constraint and thus avoid the Prokhorov metric. For instance, Beran (1977a, 1977b) first substitutes a continuous distribution to any discrete data set and, then, performs estimations by minimizing the Hellinger distance.

Except when Ω is the real axis, it is not possible to define cumulative distribution functions (cdf); thus most arguments in this work are relative to probability density functions (pdf).

Influence function

For regular functional $T(f)$ of distribution f, expansion is possible in the vicinity of f in terms of the *von Mises derivatives*. Let distribution g be in this vicinity, then

$$T(g) = T(f) + \int \Psi(\omega)\; g\;(\omega)\; d\omega + \ldots$$

Truncation at the level of the first order term is the more valid, the smaller the *Prokhorov* distance $d(f,g)$ is.

The function $\Psi(\omega)$, which depends upon the distribution f, has been named "influence" function by Hampel (1974). Its evaluation is easily obtained through

$$\begin{aligned}\Psi(\omega_0) &= \int \Psi(\omega)\; \delta(\omega-\omega_0)\; d\omega \\ &= \lim\; \{T[\,(1-t)\; f + t\delta] - T(f)\},\; t \in R,\; t \to 0\end{aligned}$$

where $\delta(\omega-\omega_0)$ is the Dirac distribution centered on $\omega = \omega_0$. It has an important role in the assessmet of *robustness*.

Jackknife technique

The so-called jackknife method deals with the estimation of the bias and with the estimation of the variance of estimators defined by suitably regular functionals. This section is only centered on justifying the derivation which leads to the ordinary as well as to the infinitesimal jackknife. For convenience of notation, we suppose a vector-value functional as well as a multidimensional sample space.

For regular functional $\underline{T}(f)$ of distribution f, expansion is possible in the vicinity of f in terms of the *von Mises derivatives*. Let distribution g be in this vicinity, then

$$\underline{T}(g) = \underline{T}(f) + \int \underline{\Psi}(\omega)\, g(\omega)\, d\omega + \frac{1}{2} \int g(\omega) \int \underline{\phi}(\omega,\omega)\, g(\omega)\, d\omega\, d\omega + \ldots$$

In the present context, f is the distribution underlying some sample X and is unknown, while g is an empirical distribution. The functional $\underline{T}(.)$ is some recipe (an arithmetic rule or possibly an algorithm) which produces an evaluation of a vector denoted $\underline{\theta}$, i.e.

$$\underset{\sim}{\theta} = \underline{T}(f),$$
$$\underline{\theta} = \underline{T}(g),$$

with

$$g = \sum w_i\, \delta(\underline{x} - \underline{x}_i)/\sum w_i$$
$$X = (\underline{x}_1, \ldots, \underline{x}_n),$$
$$w_i > 0.$$

The script $\delta(\underline{x} - \underline{x}_i)$ stands for the Dirac distribution centered on the observation $\underline{x}_i$ and to each observation $\underline{x}_i$ is associated a non-negative weight w_i. We introduce the expression of g in the expansion to obtain

$$\hat{\underline{\theta}} = \underline{\theta} + \sum \underline{\Psi}(\underline{x}_i)\, w_i/(\sum w_i) + \frac{1}{2} \sum\sum w_i\, \underline{\phi}(\underline{x}_i, \underline{x}_j)\, w_j/(\sum w_i)^2 + \ldots$$

which can also be written in the matricial notation

$$\hat{\underline{\theta}} = \underline{\theta} + \underline{\Psi w}/(\underline{1}'\, \underline{w}) + \frac{1}{2}\, \underline{w}' \circledast \phi \circledast \underline{w}/(\underline{1}'\, \underline{w})^2 + \ldots$$

where

$$\underline{w} = (w_1, \ldots, w_n)'$$
$$\underline{1}' = (1, \ldots, 1)'$$

and $\oplus$ is a script which would denote a matricial product if vector $\underline{\theta}$ were a scalar. Per definition, the k-th component is given by a bilinear form with an ordinary matricial product, i.e.,

$$[\underline{w}' \oplus \phi \oplus \underline{w}]_k = \underline{w}' \, [\phi]_k \, \underline{w}$$

where $[\phi]_k$ is a square matrix.

While searching for the origin of possible bias, it appears to be due to the quadratic term inasmuch as

$\hat{\underline{\theta}}$ is consistent with respect to $\underline{\theta}$

and

w_i is independent of $\underline{x}_i$.

That is

$$E\{\hat{\theta}\} = \underline{\theta} + \frac{1}{2} E\{\underline{w}' \oplus \phi \oplus \underline{w}\} \; / \; (\underline{1}' \; \underline{w})^2.$$

Terms of order superior to two have been neglected seeing they are neglegible with respect to the lower orders. Observe that the first order term cannot introduce any bias for *consistent estimators*.

To derive an expression for the variance of $\hat{\underline{\theta}}$, we limit ourselves to the first order term. Thus we only consider

$$\hat{\underline{\theta}} = \underline{\theta} + \Psi \; \underline{w} \; / \underline{1}' \; \underline{w}$$

and the covariance of $\hat{\underline{\theta}}$ is approximately given by

$$\mathrm{cov}(\hat{\underline{\theta}}) = E\{\Psi \underline{w} \; \underline{w}' \; \Psi'\} \; / \; (\underline{1}' \; \underline{w})^2.$$

These preliminary results being reported, we now produce the derivation of the jackknife method. It consists in a clever comparison of an estimator based on a set of weights $\underline{w}$ with other estimators based on different sets of weights.

To apply the method, we distribute the observations in g groups of size h (n = gh) and change the weights of the observations in the i-th group by some factor (1 + t). Then the weight-vector $\underline{w}$ becomes

$$\underline{w}_i = (I + tE_i) \; \underline{w}$$

where I is the identity matrix and E_i is diagonal with ones for the i-th group and zeros otherwise. It may be appropriate to note that the usual presentation is relative to a scalar estimator $\hat{\theta}$ and that $t = -1$ produces the ordinary jackknife, and small t the infinitesimal jackknife.

We denote by $\hat{\underline{\theta}}_i$ the pseudo-estimate based on the weight-vector $\underline{w}_i$, i.e.,

$$\hat{\underline{\theta}}_i = \underline{T}(X, \underline{w}_i).$$

The corresponding pseudo-value is given by

$$\tilde{\underline{\theta}}_i = [(\underline{1}' \underline{w}_i) \hat{\underline{\theta}}_i - (\underline{1}' \underline{w}) \hat{\underline{\theta}}]/t$$

and has the expansion, for small t,

$$\begin{aligned}\tilde{\underline{\theta}}_i &= (\underline{1}' E_i \underline{w}) \underline{\theta} + \Psi E_i \underline{w} \\ &+ \frac{1}{2} \frac{1}{\underline{1}'\underline{w}} [2\underline{w} \circledast \phi \circledast E_i \underline{w} + t\underline{w}' E_i \circledast \phi \circledast E_i \underline{w} \\ &\qquad - \frac{\underline{1}' E_i \underline{w}}{\underline{1}' \underline{w}} \underline{w}'_i \circledast \phi \circledast \underline{w}_i] + \ldots\end{aligned}$$

Averaging of these pseudo-values leads to the jackknife estimate

$$\tilde{\underline{\theta}} = (1/\underline{1}' \underline{w}) \textstyle\sum_i \tilde{\underline{\theta}}_i;$$

it has a relatively simple expansion when ϕ is block-diagonal and when all groups are equally weighted. Thus, under conditions

$$\textstyle\sum_i \sum_j \underline{w}' E_j \circledast \phi \circledast E_i \underline{w} = 0, \text{ for } i \neq j$$

and

$$g \, \underline{1}' E_i \underline{w} = \underline{1}' \underline{w}, \text{ for all } i,$$

we derive, for small t/g,

$$\tilde{\underline{\theta}} = \underline{\theta} + \Psi\underline{w}/(\underline{1}'\underline{w}) + \frac{1}{2} (1 + t - 2t/g) \underline{w}' \circledast \phi \circledast \underline{w}/(\underline{1}'\underline{w})^2 + \ldots$$

Comparison with the expansion of $\hat{\underline{\theta}}$ reveals that the jackknife estimate $\tilde{\underline{\theta}}$ equates the original estimator $\hat{\underline{\theta}}$ while the infinitesimal version is applied. Bias reduction occurs with the ordinary jackknife $(t = -1)$ by cancellation of the second order term. Exact cancellation of this bias producing term is given by the solution of the equation in t

$$\textstyle\sum_i (\underline{w}_i \circledast \phi \circledast \underline{w}_i/\underline{1}'\underline{w}_i) = g\underline{w}' \circledast \phi \circledast \underline{w}/\underline{1}'\underline{w}$$

or, under the same conditions, by precisely

$$t = -1$$

whatever the value of g is.

Estimation of the covariance of $\hat{\underline{\theta}}$ is obtained after investigating the covariance of the variables

$$\underline{\delta}_i = \underline{\tilde{\theta}}_i - (\underline{1}' E_i \underline{w}) \underline{\tilde{\theta}}.$$

The attitude is that the pseudo-value $\underline{\tilde{\theta}}_i$ is essentially representative of the i-th group incidence on the estimator $\hat{\underline{\theta}}$; therefore, its covariance could possibly lead to the covariance of $\hat{\underline{\theta}}$. This way will now be followed under two conditions.

$$E(\Psi E_i \underline{w}) (\underline{w}' E_j \Psi') = 0, \text{ for } i \neq j$$

and

$$(\underline{1}' \underline{w}) \sum_i (\underline{1}' E_i \underline{w}) \operatorname{cov}(\Psi E_i \underline{w}) = \sum_i (\underline{1}' E_i \underline{w})^2 \operatorname{cov}(\Psi \underline{w}).$$

They imply, for the first, that the g groups are drawn independently from a population of group variates and, for the second, that each sampled group either has the same weight $(\underline{1}' E_i \underline{w})$ or yields a contribution to the total covariance proportional to its weight.

Under these conditions, we have

$$\underline{\delta}_i = \Psi E_i \underline{w} - (\underline{1}' E_i \underline{w}/\underline{1}' \underline{w}) \Psi \underline{w}$$

$$\sum_i \operatorname{cov}(\underline{\delta}_i) = [1 - \sum_i (\underline{1}' E_i \underline{w}/\underline{1}' \underline{w})^2] \operatorname{cov}(\Psi \underline{w})$$

and

$$\operatorname{cov}(\hat{\underline{\theta}}) = \sum_i \operatorname{cov}(\underline{\delta}_i)/[(\underline{1}' \underline{w})^2 - \sum_i (\underline{1}' E_i \underline{w})^2].$$

It is convenient to state this last result in terms of the derivatives of $\underline{T}(X, \underline{w})$ with respect to $\underline{w}$. Let us define the Jacobian J such that

$$\hat{\underline{\theta}}_i = \underline{T}(X, \underline{w}_i) = \hat{\underline{\theta}} + t J E_i \underline{w},$$

then we evaluate the pseudo-estimate, the pseudo-value and, finally, the jackknife estimate

$$\underline{\tilde{\theta}} = \underline{\hat{\theta}} + J\ \underline{w} = \underline{\hat{\theta}},$$

while t is small and due to the homogeneity of $\underline{T}(X, \underline{w})$. The variate $\underline{\delta}_i$ becomes in the present context

$$\underline{\delta}_i = \underline{\tilde{\theta}}_i - (\underline{1}'\ E_i\ \underline{w})\ \underline{\tilde{\theta}}$$
$$= (\underline{1}'\ \underline{w})\ J\ E_i\ \underline{w}$$

and, therefrom,

$$\text{cov}(\underline{\hat{\theta}}) = \textstyle\sum_i \text{cov}(J\ E_i\ \underline{w})/[1 - \sum_i(\underline{1}'\ E_i\ \underline{w}/\underline{1}'\ \underline{w})^2].$$

Prokhorov metric

This metric induces over the *distribution space* a vague topology, i.e., a topology of pointwise convergence. It has been introduced by Prokhorov (1956), proposed by Hampel (1971) to assess *robustness* and has progressively retained the attention in probability theory to investigate limit theorems - e.g., see Komlós et al. (1975).

We first define the Prokhorov metric between two functions f and g over a common sample space Ω and, then, illustrate for several particular distribution functions on the real axis.

In a metric space Ω, to any closed subset $\omega \subset \Omega$ we associate an open η-neighborhood ω^η of all points at distance less than η from ω. Formally, with $\rho(.,.)$ the metric of the Ω-space,

$$\omega^\eta = \{y : \exists\, x \in \omega,\ \rho(x,y) < \eta\}.$$

Then the metric is given by the distance d(f,g) between to functions f and g, associated to measures F and G, in the following way

$$d(f,g) = \max\,\{\Pi(F,G),\ \Pi(G,F)\}$$

$$\Pi(F,G) = \inf\{\varepsilon \geq 0 : F(\omega) \leq F(\omega^\eta) + \varepsilon,$$

$$\eta = \eta(\varepsilon),\ \eta(0) = 0,\ \eta'(\varepsilon) > 0,\ \text{for all } \omega \subset \Omega\}.$$

The arbitrariness in the selection of the monotonously increasing function $\eta(\varepsilon)$ is ordinarily reduced by setting η equal to ε. This is performed although ε is a scalar whereas η has the dimension of a distance in Ω. A further simplification is possible while f and g are distribution functions; effectively, one has

$$F(\Omega) = G(\Omega) \Rightarrow \Pi(F,G) = \Pi(G,F).$$

Then the simplified Prokhorov metric can be written

$$d(f,g) = \inf\{\varepsilon \geq 0 : F(\omega) \leq G(\omega^\varepsilon) + \varepsilon,\ \text{for all } \omega \subset \Omega\}$$

or, equivalently,

$$d(f,g) = \inf\{\varepsilon \geq 0 : G(\omega) \leq F(\omega^\varepsilon) + \varepsilon,\ \text{for all } \omega \subset \Omega\}.$$

It is bounded as follows

$$0 \leq d(f,g) \leq F(\Omega) = G(\Omega) = 1.$$

To provide some insight, we consider the distances between several distribution functions defined over the real axis, $\Omega = R^1$. They are

- f : uniform in $[0,a]$, with $a > 0$
- g : uniform in $[0,b]$, with $b \geq a$
- $h = t\,f + (1 - t)g$, with $0 \leq t \leq 1$
- f_n : a sampling of size n from f.

Distribution h is a step function with a jump at coordinate $x = a$; it is a linear interpolation between f and g. Distribution f_n is a weighted sum of Dirac functions δ,

$$f_n = \frac{1}{n} \sum \delta(x - x_i),$$

where x_i are the n sampled values in the interval $[0,a]$. We will later assume that the sample has been ranked according to

$$0 < x_1 < \ldots < x_n < a.$$

This is not restrictive and can be performed with probability one.

The distance between f and h is given by

$$d(f,h) = (1 - t)\ (b - a)/(b + 1 - t)$$

and is realized by $\omega = [0,a]$ for $\Pi(F,H)$, as well as $\omega = [a + \varepsilon,\ b]$ for $\Pi(H,F)$. For $t = 0$ we obtain

$$d(f,g) = (b - a)/(b + 1).$$

Similarly, the distance between h and g is given by

$$d(h,g) = t\ (b - a)/(b + 1)$$

and the triangular inequality holds true, although h is a linear interpolation. We have

$$d(f,g) < d(f,h) + d(h,g),\ 0 < t < 1.$$

To derive the distance between f and f_n, it is convenient to make use of $\Pi(F,F_n)$; but great care is needed to define the subset ω_0 realizing the infimum. we are compelled to introduce some notations for this only derivation. The general idea of the ω_0 construction is to sum small connex intervals e_i selected in order to have

$$F_n\ (e_i^{\varepsilon}) = 0 \text{ and } F(e_i) = \max.$$

Then

$$\omega_0 = \cup\ e_i.$$

Denote by y_i all disjoint finite intervals created by the sample of size n, and by l_i their lengths; formally

$$y_i = [x_{i-1}, x_i]$$

and

$$l_i = x_i - x_{i-1}$$

with

$$x_0 = 0,\ x_{n+1} = a.$$

The selection of e_i is implicitely associated to the knowledge of ε. In each interval y_i, satisfying

$$l_i > 2\varepsilon,$$

we select the intervals

$$e_i = [x_{i-1} + \varepsilon,\ x_i - \varepsilon]$$

of lengths $(l_i - 2\varepsilon)$ and such that

$$[e_i^{\varepsilon}] = y_i.$$

Therefore, the distance definition becomes

$$\Pi(F,F_n) = \{\varepsilon : F(\omega_0) = F_n(\omega_0^{\varepsilon}) + \varepsilon\}$$

or, immediately,

$$(1/a)\ \sum s_i(l_i - 2\varepsilon) = 0 + \varepsilon$$

and

$$d(f,f_n) = \varepsilon = \sum s_i\ l_i/(a + 2 \sum s_i)$$

with

$$s_i = 1,\ \text{if } l_i > 2\varepsilon$$
$$= 0,\ \text{otherwise.}$$

The way f_n converges to f, while the size n increases, is of great interest and will thus be investigated.

We assume n sufficiently large to assimilate the sample distribution of variate l_i to its continuous parent distribution

$$\text{pdf}\,\{l\} = (n/a)\exp(-\,nl/a).$$

With this only condition, ε is solution of the implicit equation

$$e = n\int_{2\varepsilon}^{\infty} l\ \text{pdf}\,\{l\}\,dl/(a + 2n\int_{2\varepsilon}^{\infty}\text{pdf}\,\{l\}\,dl).$$

This leads to a law of iterated logarithm

$$\varepsilon = -\,[a/(2n)]\ \ln\,\varepsilon$$

and to the approximation

$$d(f,f_n) = \varepsilon \approx -\,[a/(2n)]\ \ln\,\{[a/(2n)]\ \ln\,(2n/a)\}.$$

To conclude, we observe that only a fraction ε of the intervals y_i is involved in the construction of ω_0, i.e.

$$\lim\,(1/n)\sum s_i = \varepsilon.$$

This may contribute to explain why the convergence in Prokhorov metric is so much faster than with the Kolmogorov metric.

Experimental validation of the above theoretical derivation has been obtained as indicated by the following results.

Parameter a	1	1
Sample size n	1000	10000
Theoretical $d(f,f_n)$	.00292	.000392
Average $\hat{d}(f,f_n)$	.00289	.000377
Stand. dev. $\hat{d}(f,f_n)$	.00032	.000021
Number of replicates to obtain the last two lines	50	20

Robustness

Let f, g, ϕ be distributions in some *distribution spaces* and d(.|.) be a distance function such as the *Prokhorov metric*, then we denote by $\phi(t_n,g)$ the distribution of the estimator t_n based on a sample of size n drawn from distribution g. Estimator t_n is robust with respect to g if, and only if,

$$\forall \varepsilon > 0, \exists \delta > 0, \forall f, \forall n :$$
$$\{d(f,g) < \delta \Rightarrow d[\phi(t_n,g), \phi(t_n,f)] < \varepsilon\}.$$

Hampel (1971) also defines the Π - robustness in order to relate estimators based on different sample sizes despite lack of independency between the samples.

Under some incompletely specified conditions, it is possible to relate ε and δ when t_n can be expanded in term of the first order *von Mises derivative*. We have

$$t_n(f) = t_n(g) + \int \psi(\omega)\, f(\omega)\, d\omega + \ldots$$

and the integral term is the more important the more differing f is from g at the point where $\psi(\omega)$ is the most important. Let ω_0 be this point of the sample space, i.e. such that

$$|\psi(\omega)| \leq |\psi(\omega_0)|,$$

then the least different f which produces an estimator bias

$$h = |t_n(f) - t_n(g)|$$
$$= |\int \psi(\omega)\, f(\omega)\, d\omega|$$

is the distribution

$$f(\omega) = (1 - t)\, g(\omega) + t\, \delta(\omega - \omega_0)$$

with

$$t = |h/\psi(\omega_0)|.$$

For small h, we have asymptotically ($n \to \infty$)

$$d(f,g) = t$$
$$d[\phi(t_n,g), \phi(t_n,f)] = h$$

and, thus,

$$d(f,g) \leq \delta \Rightarrow \varepsilon \geq \delta|\psi(\omega_0)|.$$

von Mises derivative

This derivative is relative to a functional of a distribution. Let $T(f)$ be the functional and f, as well as g, be distributions of some *distribution space* defined over some sample space Ω; then the first order von Mises derivative (1947) is, per definition,

$$\lim \{T[(1-t)f + tg] - T(f)\} / t = \int \Psi(\omega)\, g(\omega)\, d\omega, \quad t \in R, \; t \to 0$$

where the limit is understood for positive t tending to zero and the integral is extended over Ω.

The existence conditions for this Gateaux derivative appear to be essentially depending upon the functional T, seeing that the convexity of the distribution space has been assumed. In order to provide some insight on the mechanism which leads to the above expression, we derive a Taylor-like expansion of $T(f)$ in the vicinity of the distribution f. A by-product of this derivation will be a sufficient set of conditions.

The presentation aims at providing insight. We will see that the Taylor-like expansion is reasonable but we will remain unable to demonstrate its valitity. Considering asymptotic situations, it is possible to progress a few steps further as did Reeds (1976). However his work is not really pertinent to our needs for he is dependent upon intricate topologies.

We assume that f and g are continuous with respect to the metric $\rho(.,.)$ of the sample space Ω. If there are discontinuities, we substitute, to f and g, new distributions which are continuous, say f_0 and g_0. These new distributions will be defined with the help of some parameter c satisfying

$$\lim d(f, f_0) = 0, \quad c \in R, \; c \to c_0,$$

where $d(.,.)$ is the *Prokhorov metric*, and

$$\lim d(g, g_0) = 0, \quad c \in R, \; c \to c_0.$$

We expect $T(f)$ to fulfill the condition

$$\lim T(f_0) = T(f), \quad c \in R, \; c \to c_0 .$$

With continuous f and g, the principle of the derivation consists in performing a fine partition of the sample space Ω, then obtaining the expansion of T as a function defined in a multidimensional space and, to conclude, refine the partition in order to obtain limit expressions.

First we restrict the Ω-space to a ball of radius R centered on some point $\underline{x}_0 \in \Omega$

$$\Omega_R = \{\underline{x} : \rho(\underline{x}, \underline{x}_0) < R\}.$$

In this ball, we consider the partition in n disjoint subsets δ_i, and to each subset we associate a measure $\Delta(\omega_i)$. We have

$$\cup\, \omega_i = \Omega_R,\ 1 \leqslant i \leqslant n$$

and $$\omega_i \cap \omega_j = 0,\ i \neq j$$

as well as $$\Delta(\omega_i \cup \omega_j) = \Delta(\omega_i) + \Delta(\omega_j),\ i \neq j.$$

In each subset we arbitrarily select a point $\underline{x}_i$ and denote by f_i and g_i the corresponding values of the distributions f and g, i.e.

$$f_i = f(\underline{x}_i),\ \underline{x}_i \in \omega_i \subset \Omega_R,$$

and $$g_i = g(\underline{x}_i),\ \underline{x}_i \in \omega_i \subset \Omega_R.$$

We now investigate the behaviour of the functional T(f) on the partitioned space. Over this ball, we denote the functional by the notation $T_R(f)$, whereas the script $T_R(\{f_i\})$ denotes its approximation based on the n evaluations in the subsets ω_i. But $T_R(\{f_i\})$ can be seen as a function defined over an n-dimension space and a Taylor expansion in the vicinity of $\{f_i\}$ is valid while T is sufficiently differentiable. We are interested by the expansion

$$\begin{aligned} T_R(\{f_i + t\,(g_i - f_i)\}) &= T_R(\{f_i\}) \\ &+ \sum_i [t(g_i - f_i)]\ h_i\ \Delta(\omega_i) \\ &+ \frac{1}{2} \sum\sum [t(g_i - f_i)]\ h_{ij}\ \Delta(\omega_i)\ \Delta(\omega_j)\ [t(g_j - f_j)] \\ &+ \frac{1}{6} \sum\sum\sum \ldots . \end{aligned}$$

where $h_i\ \Delta(\omega_i)$ stands for the first partial derivative $(\partial/\partial f_i)\ T_R\{f_i\}$ and similarly for the higher order partial derivatives.

The last part of the derivation consists in obtaining the limit expressions. We first refine the partition, letting n tend to infinity with all $\Delta(\omega_i)$ tending to zero. Then

$$T_R[f + t(g - f)] = T_R(f)$$

$$+ t \int h(\omega)[g(\omega) - f(\omega)]\, d\omega$$

$$+ \frac{1}{2} t^2 \int [g(\omega) - f(\omega)] \int h(\omega,\omega)[g(\omega) - f(\omega)]\, d\omega\, d\omega$$

$$+ \frac{1}{6} t^3 \ \dots .$$

We assume existence of the limits and the integrals are understood in the Riemann-Stieltjes sense. The domain of integration is, thus far, the ball Ω_R. We now let R tend to infinity (and also possibly c tend to c_0) in order to obtain

$$T[(1 - t) f + tg] = T(f)$$

$$+ t \int h(\omega)[g(\omega) - f(\omega)]\, d\omega$$

$$+ \frac{1}{2} t^2 \int [g(\omega) - f(\omega)] \int h(\omega,\omega)[g(\omega) - f(\omega)]\, d\omega\, d\omega$$

$$+ \frac{1}{6} t^3 \ \dots$$

with the whole sample space Ω as integration domain.

To conclude we reorganize the integral expressions, as follows for the first order term, and similarly for the following ones.

$$\int h(\omega)[g(\omega) - f(\omega)]\, d\omega$$

$$= \int h(\omega) g(\omega)\, d\omega - [\int h(\omega) f(\omega)\, d\omega]$$

$$= \int \{h(\omega) - [\int h(\omega) f(\omega)\, d\omega]\} g(\omega)\, d\omega$$

$$= \int \psi(\omega) g(\omega)\, d\omega.$$

Inasmuch as T(f) is a functional sufficiently regular to permit all the passages to limits, the Taylor-like expansion

$$T[(1 - t) f + tg] =$$

$$T(f) + t \int \psi(\omega) g(\omega)\, d\omega$$

$$+ \frac{1}{2} t^2 \int g(\omega) \int \psi(\omega, \omega) g(\omega)\, d\omega\, d\omega + \dots$$

is valid and the expression of the von Mises derivative results immediately.

When we introduce $g = f$ in the two members of the above equality, we obtain a tautology for any t value, and therefore we have

$$\int \Psi(\omega)\, f(\omega)\, d\omega = 0$$

as well as

$$\int f(\omega) \int \psi(\omega, \omega)\, f(\omega)\, d\omega\, d\omega = 0 .$$

The expansion in Taylor series has been used by Filippova (1962) to obtain asymptotic properties of statistical estimators, we use it essentially in small sample contexts. It is the basis for the *jackknife* study as well as for the *influence function* concept.

AUTHOR INDEX

SUBJECT INDEX

Vol. 521: G. Cherlin, Model Theoretic Algebra – Selected Topics. IV, 234 pages. 1976.

Vol. 522: C. O. Bloom and N. D. Kazarinoff, Short Wave Radiation Problems in Inhomogeneous Media: Asymptotic Solutions. V. 104 pages. 1976.

Vol. 523: S. A. Albeverio and R. J. Høegh-Krohn, Mathematical Theory of Feynman Path Integrals. IV, 139 pages. 1976.

Vol. 524: Séminaire Pierre Lelong (Analyse) Année 1974/75. Edité par P. Lelong. V, 222 pages. 1976.

Vol. 525: Structural Stability, the Theory of Catastrophes, and Applications in the Sciences. Proceedings 1975. Edited by P. Hilton. VI, 408 pages. 1976.

Vol. 526: Probability in Banach Spaces. Proceedings 1975. Edited by A. Beck. VI, 290 pages. 1976.

Vol. 527: M. Denker, Ch. Grillenberger, and K. Sigmund, Ergodic Theory on Compact Spaces. IV, 360 pages. 1976.

Vol. 528: J. E. Humphreys, Ordinary and Modular Representations of Chevalley Groups. III, 127 pages. 1976.

Vol. 529: J. Grandell, Doubly Stochastic Poisson Processes. X, 234 pages. 1976.

Vol. 530: S. S. Gelbart, Weil's Representation and the Spectrum of the Metaplectic Group. VII, 140 pages. 1976.

Vol. 531: Y.-C. Wong, The Topology of Uniform Convergence on Order-Bounded Sets. VI, 163 pages. 1976.

Vol. 532: Théorie Ergodique. Proceedings 1973/1974. Edité par J.-P. Conze and M. S. Keane. VIII, 227 pages. 1976.

Vol. 533: F. R. Cohen, T. J. Lada, and J. P. May, The Homology of Iterated Loop Spaces. IX, 490 pages. 1976.

Vol. 534: C. Preston, Random Fields. V, 200 pages. 1976.

Vol. 535: Singularités d'Applications Differentiables. Plans-sur-Bex. 1975. Edité par O. Burlet et F. Ronga. V, 253 pages. 1976.

Vol. 536: W. M. Schmidt, Equations over Finite Fields. An Elementary Approach. IX, 267 pages. 1976.

Vol. 537: Set Theory and Hierarchy Theory. Bierutowice, Poland 1975. A Memorial Tribute to Andrzej Mostowski. Edited by W. Marek, M. Srebrny and A. Zarach. XIII, 345 pages. 1976.

Vol. 538: G. Fischer, Complex Analytic Geometry. VII, 201 pages. 1976.

Vol. 539: A. Badrikian, J. F. C. Kingman et J. Kuelbs, Ecole d'Eté de Probabilités de Saint Flour V-1975. Edité par P.-L. Hennequin. IX, 314 pages. 1976.

Vol. 540: Categorical Topology, Proceedings 1975. Edited by E. Binz and H. Herrlich. XV, 719 pages. 1976.

Vol. 541: Measure Theory, Oberwolfach 1975. Proceedings. Edited by A. Bellow and D. Kölzow. XIV, 430 pages. 1976.

Vol. 542: D. A. Edwards and H. M. Hastings, Čech and Steenrod Homotopy Theories with Applications to Geometric Topology. VII, 296 pages. 1976.

Vol. 543: Nonlinear Operators and the Calculus of Variations, Bruxelles 1975. Edited by J. P. Gossez, E. J. Lami Dozo, J. Mawhin, and L. Waelbroeck, VII, 237 pages. 1976.

Vol. 544: Robert P. Langlands, On the Functional Equations Satisfied by Eisenstein Series. VII, 337 pages. 1976.

Vol. 545: Noncommutative Ring Theory. Kent State 1975. Edited by J. H. Cozzens and F. L. Sandomierski. V, 212 pages. 1976.

Vol. 546: K. Mahler, Lectures on Transcendental Numbers. Edited and Completed by B. Diviš and W. J. Le Veque. XXI, 254 pages. 1976.

Vol. 547: A. Mukherjea and N. A. Tserpes, Measures on Topological Semigroups: Convolution Products and Random Walks. V, 197 pages. 1976.

Vol. 548: D. A. Hejhal, The Selberg Trace Formula for PSL (2,ℝ). Volume I. VI, 516 pages. 1976.

Vol. 549: Brauer Groups, Evanston 1975. Proceedings. Edited by D. Zelinsky. V, 187 pages. 1976.

Vol. 550: Proceedings of the Third Japan – USSR Symposium on Probability Theory. Edited by G. Maruyama and J. V. Prokhorov. VI, 722 pages. 1976.

Vol. 551: Algebraic K-Theory, Evanston 1976. Proceedings. Edited by M. R. Stein. XI, 409 pages. 1976.

Vol. 552: C. G. Gibson, K. Wirthmüller, A. A. du Plessis and E. J. N. Looijenga. Topological Stability of Smooth Mappings. V, 155 pages. 1976.

Vol. 553: M. Petrich, Categories of Algebraic Systems. Vector and Projective Spaces, Semigroups, Rings and Lattices. VIII, 217 pages. 1976.

Vol. 554: J. D. H. Smith, Mal'cev Varieties. VIII, 158 pages. 1976.

Vol. 555: M. Ishida, The Genus Fields of Algebraic Number Fields. VII, 116 pages. 1976.

Vol. 556: Approximation Theory. Bonn 1976. Proceedings. Edited by R. Schaback and K. Scherer. VII, 466 pages. 1976.

Vol. 557: W. Iberkleid and T. Petrie, Smooth S^1 Manifolds. III, 163 pages. 1976.

Vol. 558: B. Weisfeiler, On Construction and Identification of Graphs. XIV, 237 pages. 1976.

Vol. 559: J.-P. Caubet, Le Mouvement Brownien Relativiste. IX, 212 pages. 1976.

Vol. 560: Combinatorial Mathematics, IV, Proceedings 1975. Edited by L. R. A. Casse and W. D. Wallis. VII, 249 pages. 1976.

Vol. 561: Function Theoretic Methods for Partial Differential Equations. Darmstadt 1976. Proceedings. Edited by V. E. Meister, N. Weck and W. L. Wendland. XVIII, 520 pages. 1976.

Vol. 562: R. W. Goodman, Nilpotent Lie Groups: Structure and Applications to Analysis. X, 210 pages. 1976.

Vol. 563: Séminaire de Théorie du Potentiel. Paris, No. 2. Proceedings 1975–1976. Edited by F. Hirsch and G. Mokobodzki. VI, 292 pages. 1976.

Vol. 564: Ordinary and Partial Differential Equations, Dundee 1976. Proceedings. Edited by W. N. Everitt and B. D. Sleeman. XVIII, 551 pages. 1976.

Vol. 565: Turbulence and Navier Stokes Equations. Proceedings 1975. Edited by R. Temam. IX, 194 pages. 1976.

Vol. 566: Empirical Distributions and Processes. Oberwolfach 1976. Proceedings. Edited by P. Gaenssler and P. Révész. VII, 146 pages. 1976.

Vol. 567: Séminaire Bourbaki vol. 1975/76. Exposés 471–488. IV, 303 pages. 1977.

Vol. 568: R. E. Gaines and J. L. Mawhin, Coincidence Degree, and Nonlinear Differential Equations. V, 262 pages. 1977.

Vol. 569: Cohomologie Etale SGA 4½. Séminaire de Géométrie Algébrique du Bois-Marie. Edité par P. Deligne. V, 312 pages. 1977.

Vol. 570: Differential Geometrical Methods in Mathematical Physics, Bonn 1975. Proceedings. Edited by K. Bleuler and A. Reetz. VIII, 576 pages. 1977.

Vol. 571: Constructive Theory of Functions of Several Variables, Oberwolfach 1976. Proceedings. Edited by W. Schempp and K. Zeller. VI. 290 pages. 1977

Vol. 572: Sparse Matrix Techniques, Copenhagen 1976. Edited by V. A. Barker. V, 184 pages. 1977.

Vol. 573: Group Theory, Canberra 1975. Proceedings. Edited by R. A. Bryce, J. Cossey and M. F. Newman. VII, 146 pages. 1977.

Vol. 574: J. Moldestad, Computations in Higher Types. IV, 203 pages. 1977.

Vol. 575: K-Theory and Operator Algebras, Athens, Georgia 1975. Edited by B. B. Morrel and I. M. Singer. VI, 191 pages. 1977.

Vol. 576: V. S. Varadarajan, Harmonic Analysis on Real Reductive Groups. VI, 521 pages. 1977.

Vol. 577: J. P. May, E_∞ Ring Spaces and E_∞ Ring Spectra. IV, 268 pages. 1977.

Vol. 578: Séminaire Pierre Lelong (Analyse) Année 1975/76. Edité par P. Lelong. VI, 327 pages. 1977.

Vol. 579: Combinatoire et Représentation du Groupe Symétrique, Strasbourg 1976. Proceedings 1976. Edité par D. Foata. IV, 339 pages. 1977.

Vol. 580: C. Castaing and M. Valadier, Convex Analysis and Measurable Multifunctions. VIII, 278 pages. 1977.

Vol. 581: Séminaire de Probabilités XI, Université de Strasbourg. Proceedings 1975/1976. Edité par C. Dellacherie, P. A. Meyer et M. Weil. VI, 574 pages. 1977.

Vol. 582: J. M. G. Fell, Induced Representations and Banach *-Algebraic Bundles. IV, 349 pages. 1977.

Vol. 583: W. Hirsch, C. C. Pugh and M. Shub, Invariant Manifolds. IV, 149 pages. 1977.

Vol. 584: C. Brezinski, Accélération de la Convergence en Analyse Numérique. IV, 313 pages. 1977.

Vol. 585: T. A. Springer, Invariant Theory. VI, 112 pages. 1977.

Vol. 586: Séminaire d'Algèbre Paul Dubreil, Paris 1975-1976 (29ème Année). Edited by M. P. Malliavin. VI, 188 pages. 1977.

Vol. 587: Non-Commutative Harmonic Analysis. Proceedings 1976. Edited by J. Carmona and M. Vergne. IV, 240 pages. 1977.

Vol. 588: P. Molino, Théorie des G-Structures: Le Problème d'Equivalence. VI, 163 pages. 1977.

Vol. 589: Cohomologie l-adique et Fonctions L. Séminaire de Géométrie Algébrique du Bois-Marie 1965-66, SGA 5. Edité par L. Illusie. XII, 484 pages. 1977.

Vol. 590: H. Matsumoto, Analyse Harmonique dans les Systèmes de Tits Bornologiques de Type Affine. IV, 219 pages. 1977.

Vol. 591: G. A. Anderson, Surgery with Coefficients. VIII, 157 pages. 1977.

Vol. 592: D. Voigt, Induzierte Darstellungen in der Theorie der endlichen, algebraischen Gruppen. V, 413 Seiten. 1977.

Vol. 593: K. Barbey and H. König, Abstract Analytic Function Theory and Hardy Algebras. VIII, 260 pages. 1977.

Vol. 594: Singular Perturbations and Boundary Layer Theory, Lyon 1976. Edited by C. M. Brauner, B. Gay, and J. Mathieu. VIII, 539 pages. 1977.

Vol. 595: W. Hazod, Stetige Faltungshalbgruppen von Wahrscheinlichkeitsmaßen und erzeugende Distributionen. XIII, 157 Seiten. 1977.

Vol. 596: K. Deimling, Ordinary Differential Equations in Banach Spaces. VI, 137 pages. 1977.

Vol. 597: Geometry and Topology, Rio de Janeiro, July 1976. Proceedings. Edited by J. Palis and M. do Carmo. VI, 866 pages. 1977.

Vol. 598: J. Hoffmann-Jørgensen, T. M. Liggett et J. Neveu, Ecole d'Eté de Probabilités de Saint-Flour VI – 1976. Edité par P.-L. Hennequin. XII, 447 pages. 1977.

Vol. 599: Complex Analysis, Kentucky 1976. Proceedings. Edited by J. D. Buckholtz and T. J. Suffridge. X, 159 pages. 1977.

Vol. 600: W. Stoll, Value Distribution on Parabolic Spaces. VIII, 216 pages. 1977.

Vol. 601: Modular Functions of one Variable V, Bonn 1976. Proceedings. Edited by J.-P. Serre and D. B. Zagier. VI, 294 pages. 1977.

Vol. 602: J. P. Brezin, Harmonic Analysis on Compact Solvmanifolds. VIII, 179 pages. 1977.

Vol. 603: B. Moishezon, Complex Surfaces and Connected Sums of Complex Projective Planes. IV, 234 pages. 1977.

Vol. 604: Banach Spaces of Analytic Functions, Kent, Ohio 1976. Proceedings. Edited by J. Baker, C. Cleaver and Joseph Diestel. VI, 141 pages. 1977.

Vol. 605: Sario et al., Classification Theory of Riemannian Manifolds. XX, 498 pages. 1977.

Vol. 606: Mathematical Aspects of Finite Element Methods. Proceedings 1975. Edited by I. Galligani and E. Magenes. VI, 362 pages. 1977.

Vol. 607: M. Métivier, Reelle und Vektorwertige Quasimartingale und die Theorie der Stochastischen Integration. X, 310 Seiten. 1977.

Vol. 608: Bigard et al., Groupes et Anneaux Réticulés. XIV, 334 pages. 1977.

Vol. 609: General Topology and Its Relations to Modern Analysis and Algebra IV. Proceedings 1976. Edited by J. Novák. XVIII, 225 pages. 1977.

Vol. 610: G. Jensen, Higher Order Contact of Submanifolds of Homogeneous Spaces. XII, 154 pages. 1977.

Vol. 611: M. Makkai and G. E. Reyes, First Order Categorical Logic. VIII, 301 pages. 1977.

Vol. 612: E. M. Kleinberg, Infinitary Combinatorics and the Axiom of Determinateness. VIII, 150 pages. 1977.

Vol. 613: E. Behrends et al., L^p-Structure in Real Banach Spaces. X, 108 pages. 1977.

Vol. 614: H. Yanagihara, Theory of Hopf Algebras Attached to Group Schemes. VIII, 308 pages. 1977.

Vol. 615: Turbulence Seminar, Proceedings 1976/77. Edited by P. Bernard and T. Ratiu. VI, 155 pages. 1977.

Vol. 616: Abelian Group Theory, 2nd New Mexico State University Conference, 1976. Proceedings. Edited by D. Arnold, R. Hunter and E. Walker. X, 423 pages. 1977.

Vol. 617: K. J. Devlin, The Axiom of Constructibility: A Guide for the Mathematician. VIII, 96 pages. 1977.

Vol. 618: I. I. Hirschman, Jr. and D. E. Hughes, Extreme Eigen Values of Toeplitz Operators. VI, 145 pages. 1977.

Vol. 619: Set Theory and Hierarchy Theory V, Bierutowice 1976. Edited by A. Lachlan, M. Srebrny, and A. Zarach. VIII, 358 pages. 1977.

Vol. 620: H. Popp, Moduli Theory and Classification Theory of Algebraic Varieties. VIII, 189 pages. 1977.

Vol. 621: Kauffman et al., The Deficiency Index Problem. VI, 112 pages. 1977.

Vol. 622: Combinatorial Mathematics V, Melbourne 1976. Proceedings. Edited by C. Little. VIII, 213 pages. 1977.

Vol. 623: I. Erdelyi and R. Lange, Spectral Decompositions on Banach Spaces. VIII, 122 pages. 1977.

Vol. 624: Y. Guivarc'h et al., Marches Aléatoires sur les Groupes de Lie. VIII, 292 pages. 1977.

Vol. 625: J. P. Alexander et al., Odd Order Group Actions and Witt Classification of Innerproducts. IV, 202 pages. 1977.

Vol. 626: Number Theory Day, New York 1976. Proceedings. Edited by M. B. Nathanson. VI, 241 pages. 1977.

Vol. 627: Modular Functions of One Variable VI, Bonn 1976. Proceedings. Edited by J.-P. Serre and D. B. Zagier. VI, 339 pages. 1977.

Vol. 628: H. J. Baues, Obstruction Theory on the Homotopy Classification of Maps. XII, 387 pages. 1977.

Vol. 629: W. A. Coppel, Dichotomies in Stability Theory. VI, 98 pages. 1978.

Vol. 630: Numerical Analysis, Proceedings, Biennial Conference, Dundee 1977. Edited by G. A. Watson. XII, 199 pages. 1978.

Vol. 631: Numerical Treatment of Differential Equations. Proceedings 1976. Edited by R. Bulirsch, R. D. Grigorieff, and J. Schröder. X, 219 pages. 1978.

Vol. 632: J.-F. Boutot, Schéma de Picard Local. X, 165 pages. 1978.

Vol. 633: N. R. Coleff and M. E. Herrera, Les Courants Résiduels Associés à une Forme Méromorphe. X, 211 pages. 1978.

Vol. 634: H. Kurke et al., Die Approximationseigenschaft lokaler Ringe. IV, 204 Seiten. 1978.

Vol. 635: T. Y. Lam, Serre's Conjecture. XVI, 227 pages. 1978.

Vol. 636: Journées de Statistique des Processus Stochastiques, Grenoble 1977, Proceedings. Edité par Didier Dacunha-Castelle et Bernard Van Cutsem. VII, 202 pages. 1978.

Vol. 637: W. B. Jurkat, Meromorphe Differentialgleichungen. VII, 194 Seiten. 1978.

Vol. 638: P. Shanahan, The Atiyah-Singer Index Theorem, An Introduction. V, 224 pages. 1978.

Vol. 639: N. Adasch et al., Topological Vector Spaces. V, 125 pages. 1978.